NEUTRON RADIOGRAPHY HANDBOOK

NEUTRON RADIOGRAPHY HANDBOOK

EUR 7622e **1981**

Abstract

The handbook has been prepared by the Euratom Working Group on Neutron Radiography (NRWG).
In the introductory chapter the principles of neutron radiography and its practical application for non-destructive testing of materials are outlined.
The next chapter gives, in detail, recommendations for carrying out neutron radiographic examinations of materials and components.
Special indicators which have been developed to test image quality and accuracy of dimension measurements taken from neutron radiographs of nuclear fuel are described in a third chapter.
A collection of selected neutron radiographs of nuclear fuel is presented in chapter four, to allow classification of observed typical defects, which are revealed by neutron radiography.
In the last chapter of the handbook technical data on the Neutron Radiography Installations in the European Community have been compiled in tables and schematic figures.

Commission of the European Communities

NEUTRON RADIOGRAPHY HANDBOOK

Nuclear Science and Technology

Edited by

P. VON DER HARDT and H. RÖTTGER

C.E.C., Joint Research Centre, Petten Establishment

D. REIDEL PUBLISHING COMPANY

DORDRECHT : HOLLAND / BOSTON : U.S.A.

LONDON : ENGLAND

Library of Congress Cataloging in Publication Data

DATA APPEAR ON SEPARATE CARD.

CIP

ISBN-13: 978-94-009-8569-8 e-ISBN-13: 978-94-009-8567-4
DOI:10.1007/978-94-009-8567-4

Publication arrangements by
Commission of the European Communities
Directorate-General Information Market and Innovation, Luxembourg

Design frontcover: J. Wells, J.R.C. ISPRA
Lay-out: Reproduction service J.R.C. PETTEN

EUR 7622e

Published by D. Reidel Publishing Company
P.O. Box 17, 3300 AA Dordrecht, Holland

Sold and distributed in the U.S.A. and Canada
by Kluwer Boston Inc.,
190 Old Derby Street, Hingham, MA 02043, U.S.A.

In all other countries, sold and distributed
by Kluwer Academic Publishers Group,
P.O. Box 322, 3300 AH Dordrecht, Holland

D. Reidel Publishing Company is a member of the Kluwer Group

INTRODUCTION

Radiography with neutrons can yield important information not obtainable by more traditional methods. In contrast to X-rays as the major tool of visual non-destructive testing, neutrons can

- be attenuated by light materials like water, hydrocarbons, boron,
- penetrate through heavy materials like steel, lead, uranium,
- distinguish between different isotopes of certain elements,
- supply high quality radiographs of highly radioactive components.

These advantages have led to multiple applications of neutron radiography since 1955, both for non-nuclear and nuclear problems of quality assurance. The required neutron beams originate from radioisotopic sources, accelerator targets, or research reactors. Energy "tailoring" which strongly influences the interaction with certain materials adds to the versatility of the method.

Since about 1970 norms and standards have been introduced and reviewed both in Europe (Birmingham, September 1973) and the United States (Gaithersburg, February 1975). The first world conference on neutron radiography will take place in December 1981, in San Diego, U.S.A. .
In Europe the interested laboratories inside the European Community have entered into systematic collaboration through the Neutron Radiography Working Group (NRWG), since May 1979. This Handbook has been compiled as one of the common tasks undertaken by the Group. Its principal authors are

J.C. Domanus (Risø National Laboratory), and
R.S. Matfield (Joint Research Centre, Ispra)

Major contributions have been received from

R. Barbalat (CEN Saclay),
P. Boyer, (CEN Cadarache)
P. Gade-Nielsen (Risø Nat. Lab.)
L. Greim (GKSS Geesthacht)
M. Houëlle (CEA Valduc)
A. Jegat (CEN Grenoble)
A. Laporte (CEN Saclay)
H.P. Leeflang (ECN Petten)
R. Liesenborgs (SCK/CEN Mol)
F. Michel (CEN Grenoble)
R.W. Parish (AERE Harwell)
G. Rottenberg (SCK/CEN Mol)
I. Ruyter (KFA Jülich)
V. Stulens (SCK/CEN Mol)
D.J. Taylor (AERE Harwell)
H. Tourwé (SCK/CEN Mol)
G. Trezza (CNEN-CSN Casaccia)

This Handbook documents the availability, not only of a large number of research reactor-based facilities in the Community, but also of advanced equipment and solid expertise for the interpretation of neutron radiographs, serving present and future needs of Europe's industry.

P. von der Hardt
H. Röttger

TABLE OF CONTENTS

1. PRINCIPLES AND PRACTICE OF NEUTRON RADIOGRAPHY

This part of the Handbook is about neutrons, radiography, and the technique that has been developed to bring them together. It is written in three chapters, a description of the subject for the assistance of the clients of neutron radiography services; a discussion on the problems facing the designer of neutron radiography equipment and a description of some of the applications. The special terms used are explained in Appendix 1.1.

1.1. INTRODUCTION TO NEUTRON RADIOGRAPHY

1.1.1 Historical

Historically, radiography came first in 1895 with the discovery by Röntgen of a radiation which he called X-rays. He rapidly realised the technical implications and in the same year took an X-ray 'photograph' of a weld in a zinc plate. The significance of X-rays for the detection of unseen flaws was immediately seen by other workers, and experimental X-radiographs were soon produced in laboratories in Europe and the U.S.A. It was later found that the attenuation of X-rays increased smoothly with atomic number, indicating that the X-rays interacted with the orbital electrons around the atomic nucleus.
The discovery of the neutron is credited to Chadwick who, in 1932, related and hypothesised on the work of Bothe, Becker, Curie and others and assumed that the penetrating radiation produced by bombarding beryllium with alpha particles was neither positively nor negatively charged; so he called it the neutron (from Latin neuter meaning neither). He had indentified a particle which, together with the proton, was one of the basic building bricks of matter.
The radiographic applications for neutrons were not acted upon quite so rapidly as had occured with X-rays and several years intervened before the first neutron radiography experiments were started in Berlin by Kallman and Kuhn [Ref. 1]. They started work in 1935 with a small accelerator source, said to be equivalent to a 2-3 gramme Ra-Be source, and they defined the basic principles of neutron radiography and recorded them on a large number of patents filed over the next ten years or so.
The publication of their work was delayed by the second World War and it was not until 1947 that they revealed the thoroughnes of their investigations by describing most of the basic techniques in use today. They suffered the disappointment of being preceded by Peters [Ref. 2] who published the results of similar experiments in 1946.
The next development had to await the advent of nuclear reactors, and the first reactor neutron radiographs were produced in 1956 by Thewlis and Derbyshire [Ref. 3] at Harwell. They carried out their work with the BEPO reactor, and its intense neutron beam allowed them to produce radiographs of much better quality than those of Kallman and Peters. They also demonstrated the

applications of neutron radiography to specific problems by showing the flaws in a uranium cylinder, a defect in a piece of boral (boron-aluminium sandwich) and the fine structure of plant tissue. The technique developed slowly for several years until problems associated with the radiography of radioactive materials encouraged its more active revival. Several researchers reported their work in the early 1960's. But it was principally the work of Berger [Ref. 4] of Argonne Laboratories in the U.S.A., followed by Barton [Ref. 5] at Birmingham University that led to its revival.
Interest expanded rapidly and Krolick [Ref. 6] et al reported in 1968 that there were 33 centres throughout six different countries all active in neutron radiography. At that time there were 46 reactor facilities in use, three accelerators and above five isotopic sources in use of being built.
The situation is much the same today in that the reactor sources predominate, and there are still very few accelerator or isotopic sources. The number of active centres however, is now probably over 50.

1.1.2. Basic Concepts

All material objects are formed from a substance which we call matter. This is an arrangement of atoms which can take many forms varying from the regular pattern of a crystal lattice to the free moving single atoms within a gas plasma. No one has ever seen an atom – although the electron microscope allows us to get very close to seeing it – and modern theory represents it as a tiny nucleus surrounded by a diffuse cloud of electrons, the outer boundary of which is not clearly defined and may not even be spherical. The nucleus is itself a group of closely bound neutrons and protons, the overall diameter of which is some 10,000 times smaller than the size of the atom. For our purposes we will imagine the atom as consisting of an extremely small, extremely dense, nucleus surrounded by an enormous empty space (on the nuclear scale) in which a retinue of electrons maintain their regular orbital motions.
The radiographic process requires free neutrons and so they must be dislodged from the nucleus. This is achieved by bombarding the nucleus and causing it to change into smaller nuclei and a number of free neutrons. These liberated neutrons are electrically neutral (i.e. no charge) and so are able to pass through the electron cloud surrounding an atom without disturbing interactions.

Unlike the X-ray which interacts with the electron cloud, the neutron interaction is not characterised by a rational dependence on the atomic number of the object, the relationship between the two being quite random. There are practically no generalisations that can be made which relate neutron characteristics to atomic mass or atomic number, and each interaction of a neutron with an atom of a particular nuclide is unique, the nature of that reaction being only related to the energy of the neutron.
To produce a neutron radiograph we must have a continuous supply of free neutrons, and these must be directed onto the object to be radiographed. This object will modify the neutron beam by scattering or absorbing the radiation, and the beam reaching the detector will have an intensity pattern representative of the structure of the object.

1.1.3. Neutron Sources

Neutrons are produced in three ways: from an accelator, a radioisotope, or a nuclear reactor. In each case they are removed from an atom by a nuclear transmutation process and they emerge over the enormous energy range of 10^{13} electron-volts, that is from 10^{-4} to 10^{9} eV. The energy of most interest for neutron radiography is about 0,03 eV, for it is at this energy that the detectors used for neutron radiography are usually most efficient, except where the resonance characteristics of the detector foil can be utilized (see 1.1.9).

1.1.3.1 Accelerators

This is a general name given to machines that accelerate a beam of charged particles (protons, deutrons, alphas etc.) and directs them onto a target (see Fig. 1.1). An interaction takes place between the bombarding particles and the

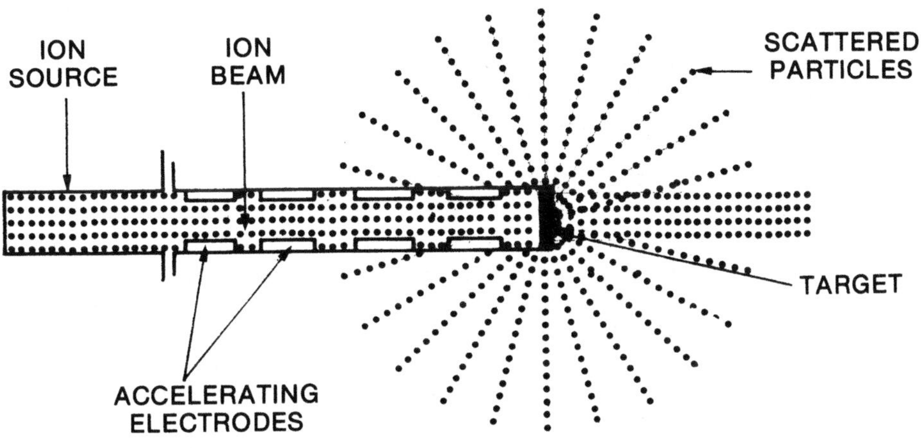

Fig. 1.1 The Principles of a Particle Accelerator.

target atoms, and this results in the expulsion of other particles. With particular combinations of incident particle and target material the ejected particles are neutrons. To remove a neutron from a target atom the energy of the bombarding particle must exceed the nuclear potential barrier surrounding the nucleus. This energy varies with both the target material and the charge on the bombarding particle, and so the target used in a particular type of accelerator is matched to the energy of incident particle that the machine can produce.
Typical of this system is the machine which uses a Penning ion source to ionise the atoms of deuterium gas and uses a Cockcroft-Walton generator (100-400 kV) to accelerate them onto a tritium target.
The reaction

$$^{3}_{1}T + ^{2}_{1}D \rightarrow ^{4}_{2}He + ^{1}_{0}n + 17{,}6 \text{ MeV}$$

takes place, that is a deuteron (^2_1D) strikes tritium (^3_1T) which releases a 14.6 MeV neutron (^1_0n), and is converted to helium (^4_2He) with a contribution of 3.6 MeV. When higher potentials are available, such as those from a Van der Graaff generator, then a beryllium or lithium target is used, and the reactions are

$$^9_4Be + ^2_1D \rightarrow ^{10}_5B + ^1_0n + 4{,}3 \text{ MeV or}$$
$$^7_3Li + ^1_1H \text{ (proton)} \longrightarrow ^7_3Be + ^1_0n + 2{,}1 \text{ MeV.}$$

An alternative system is to accelerate electrons onto a tungsten target and thereby produce X-rays. If these are directed onto a second target with a high (X,n) reaction cross-section, such as beryllium or uranium, again neutrons will be produced.
This last system has the potential of being used as a dual purpose generator of both X-rays and neutrons. Two such machines have been reported, the first is a 5,5 MeV Linac which was built as an X-ray machine [Ref. 7] and then modified to produce neutrons, and the second is a large 20 MeV Linac [Ref. 8] which was designed as a dual purpose X-ray/neutron generator. The first machine used a tungsten target, and the X-ray and gamma ray emission from this produces a $^9_4Be + \gamma \rightarrow ^8_4Be + ^1_0n$ reaction in the beryllium. The second interchanges a tungsten and uranium target, the first producing X-rays and the second generating neutrons by the reaction

$$^{238}_{92}U + \gamma \rightarrow ^{237}_{92}U + ^1_0n$$

The life and output of the target used in accelerators varies with the system, and the energy of the bombarding particle. Fig. 1.2 shows this variation of neutron yield with energy for the deuterium-tritium and the deuterium-beryllium reactions (usually referred to as 'DT' and 'DB'). The beryllium target is used in the form of pure metal, and, providing it is adequately cooled, will not deteriorate significantly with use.
Tritium targets are produced by absorbing tritium gas in a titanium or zirconium layer on a copper plate. The neutron output is high but the lifetime (usually defined as the time required for the neutron output to fall to half its initial value) is relatively low. The early machines of this type used a continuously pumped vacuum system in which the tritium is fed to the target through a controlled leak. An alternative system used a large rotating target which increased the lifetime by simply providing a larger target area. There is considerable variation in the reported life of these systems but times between 10-100 hours are usually quoted.
Later designs use a sealed accelerator tube in which the problem of the depletion of the tritium in the target was overcome by feeding a mixture of deuterium and tritium into the ion source. Tritons as well as deuterons are accelerated into the target so that the net amount of tritium in the target remains about the same, and hence the neutron yield is reasonably constant.
More detailed descriptions of the various types of particle accelerator are given in the reviews by Olive et al [Ref. 9], Krolick et al [Ref. 6] and Holland and Hawkesworth [Ref. 10], and details of source accelerator systems are given in Table 1.1

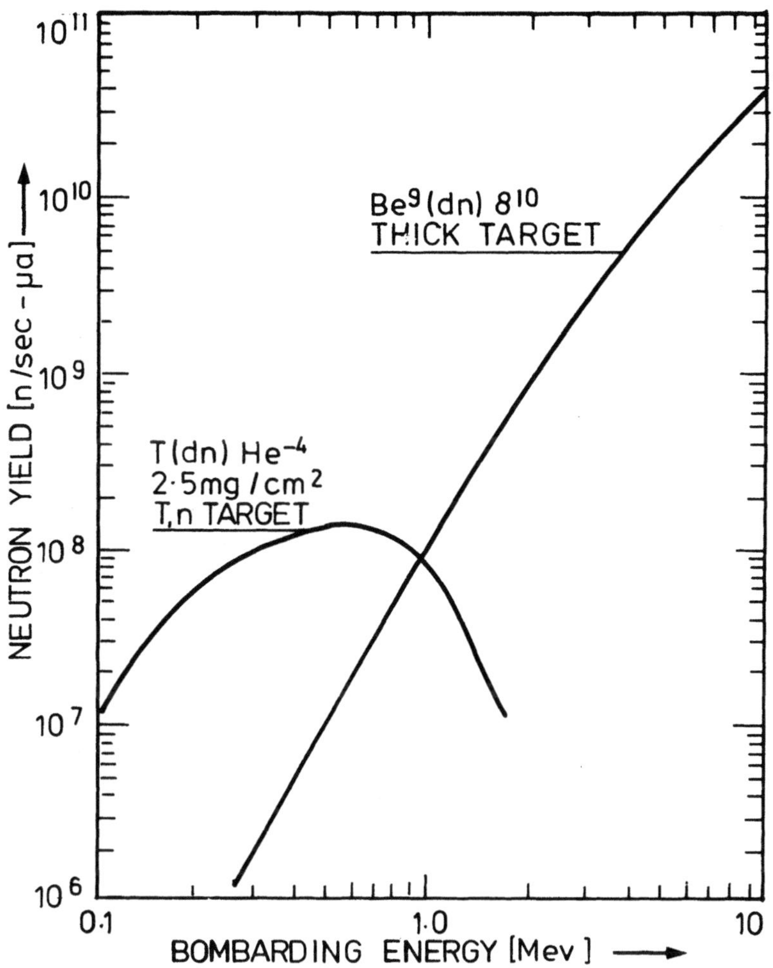

Fig. 1.2 Neutron Yield from Deuteron Reactions (After Hawkesworth [Ref. 11]).

Table 1.1 **Accelerator Neutron Sources**

Manufacturer	Type	Particle	Targets materials	Operating Voltage, kV	Beam current mA	Fast neutron output, n s^{-1}	Neutron energy MeV
Elliot Automation	P Tube	Deuteron	Tritium	120	2	10^{11}	14
20th Century Electronics	NGH 150	Deuteron	Tritium	150	3.5	4×10^{11}	4
Sames	T	Deuteron	Tritium	400	3	10^{11}	2
High Voltage Eng. Co.	Van der Graaff	Deuteron	Beryllium	3,000	0.6	[1])	1.6
Mulfard	Linac	Electron	Beryllium	5,500	0.2	2×10^{11}	1.4

1 Peak thermal flux in water 10^{10} n $cm^{-2}s^{-1}$

1.1.3.2 Radioisotopes

Radioisotopes are produced by bombarding nuclei with charged particles in an accelerator or a nuclear reactor. A nucleus becomes radioactive when it changes from a stable, unexcited, state to an unstable, excited, condition. Now, for a nucleus to be stable it must contain a particular neutron-proton ratio. This ratio varies from 1 to about 1.6 (excluding hydrogen) as the atomic number increases. The stable condition is referred to as the ground state, and if extra energy can be imparted to the nucleus it is said to have been raised to an excited state, from which it eventually decays back to the ground state, usually with the emission of gamma rays. In the excited condition there is no change in the neutron-proton ratio unless the energy imparted to the nucleus sufficiently exceeds the energy that binds it together for it to eject one of its neutrons or protons. The nucleus then become unstable because it has the wrong neutron-proton ratio for its particular atomic number.
So, by bombarding atoms with charged particles of sufficient energy it is possible to raise the nucleus to a state of instability from which it will decay back to its stable state at a characteristic rate measured by the half-line (the time take for the radioactivity to halve). Unfortunately there are few radiosotopes which emit neutrons, and neutron production is achieved in the same manner as with accelerators, that is by allowing the gamma rays or alpha particles emitted from the radioactive isotope to bombard a neutron emitting target.
The disadvantage of the radioisotope is that the activity is continuously reducing. When the radioisotope has a long half-life this is not inconvenient, but a radioisotope such as antimony loses half of its activity every 60 days and must be regularly reactivated. This, of course, adds to the cost of the neutron generator.
Beryllium has the lowest neutron binding energy (1.6 keV) of all the nuclides, and is used as a target with both alpha- and gamma-emitting radioisotopes (see Fig. 1.3). The neutrons produced from these reactions vary in energy up to a

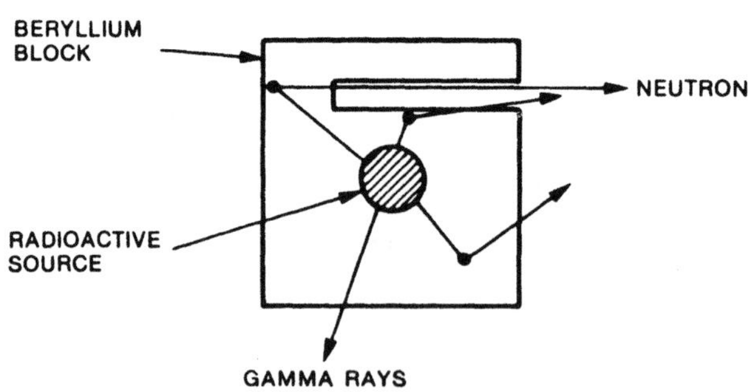

POLONIUM (Po^{210}) $\xrightarrow{\alpha}$ BERYLLIUM $\xrightarrow{n}$ 11 MeV

ANTIMONY (Sb^{124}) $\xrightarrow{\gamma}$ BERYLLIUM $\xrightarrow{n}$ 25 KeV

Fig. 1.3 Isotopic Neutron Sources.

maximum of about 11 MeV; the lowest energy ,and as we shall see later the most useful, coming from a combination of antimony and beryllium. This is an (γ,n) source and like all gamma sources has the disadvantage of requiring a lead shield to prevent the gamma rays from causing a health hazard.
There are a few radioisotopes which decay by spontaneous fission (a process described in the next section) but of these only californium-252 has sufficient neutron output to be considered here. At the time of writing, the only available supply of this material is from nuclear reactors in the U.S.A.*), and because it takes a long time to produce usable quantities it is very expensive. Fortunately, the price is falling and so in future it may be an attractive neutron source. Table 1.2 gives details of radioisotopic sources.

Table 1.2 **Radioisotopic Neutron Sources**

Source	Half-life		Reaction	Neutron yield ($n.s^{-1}\ g^{-1}$)	Neutron energy (MeV)
^{124}Sb-Be	60	d	(γ.n)	$2.7.10^{9}$	0,024
^{210}Po-Be	138	d	(α.n)	$1.28.10^{10}$	4,3
^{241}Am-Be	458	a	(α.n)	1.10^{7}	$\sim$ 4
^{226}Ra-Be	1620	a	(α.n)	$1.3.10^{7}$	$\sim$ 4
^{227}Ac-Be	21,8	a	(α.n)	$1.1.10^{9}$	$\sim$ 4
^{228}Th-Be	1,91	a	(α.n)	$1.7.10^{10}$	$\sim$ 4
^{252}Cf	2,65	a	fission	$2.34.10^{12}$	2,3

1.1.3.3 Thermal Nuclear Reactors

At the present stage of neutron radiographic development the nuclear reactor provides the most intense neutron beams and therefore the highest quality neutron radiographs. Whilst accelerators and isotopic sources are limited to a neutron flux at the detector foil of about 10^{6} n $cm^{-2}s^{-1}$ the nuclear reactor can provide a neutron flux of up to 10^{8}–10^{9} n $cm^{-2}s^{-1}$ for a comparable collimator arrangement. The disadvantage of the nuclear reactor is its lack of mobility, high capital cost and the necessity to obtain a licence to operate. Its advantage lies in its intense neutron source strength, its low cost per neutron (about 20-25 times less than an accelerator) and its lower moderation factor (see below). Most of the reactors in use for neutron radiography are principally used for nuclear research, and their resulting high utilization justifies the capital cost. The average neutron radiography facility could rarely make use of more than 20% of the neutrons available from typical nuclear reactors and so the use of reactor sources will probably be limited to organizations that can use the surplus neutrons for activation analysis, neutron physics studies, isotope production, nuclear research etc.
The nuclear reactor is an assembly in which a fissionable material, such as uranium, is dispersed in a moderating material, such as heavy water, and these are contained in a concrete radiation shield (see Fig. 1.4). Some form of cooling

*) and perhaps the USSR

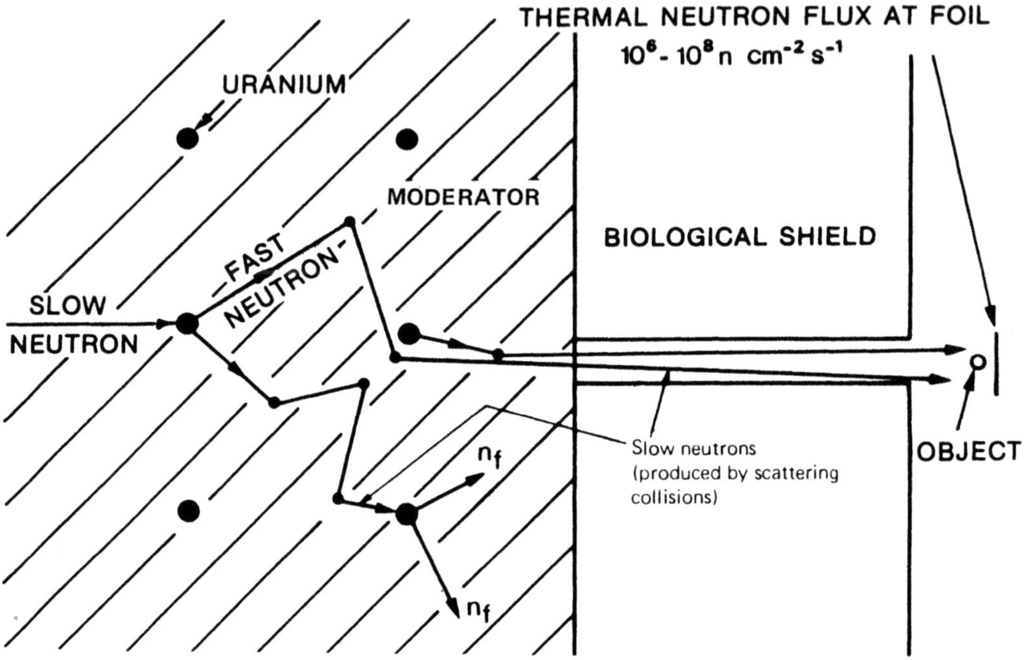

Fig. 1.4 Thermal Nuclear Reactor Source.

is provided to remove the process heat and a number of control elements are inserted into the assembly to regulate the nuclear reaction.

The fission process is induced by a neutron striking a uranium atom and thereby causing the nucleus to split into two roughly equal parts. These parts are called fission fragments and are accompanied by charged particles, gamma rays, and other neutrons. These other neutrons are available to continue the reaction by striking other nuclei and so producing further fissions in a chain reaction.

One important condition must be achieved in order to maintain this state of self-perpetuation: the liberated neutron must be slowed down in order to give it a high chance of causing further fission. This slowing down is achieved by making the fast neutron pass through an essentially non-absorbing moderating material before it hits another uranium atom. Moderating materials contain light elements such as hydrogen, carbon and beryllium, and the neutron loses energy by a series of scattering collisions, in the manner of billiard balls striking each other. For efficient neutron production the number of neutrons lost during the moderation phase must be kept as low as possible, and the uranium and moderator 'mix' in a nuclear reactor is designed to achieve this. Reactor neutrons are born at about 2 MeV and are slowed down by the moderating material to about 0.03 eV (the so-called thermal energy). This is the energy at which the neutron is in thermal equilibrium with its surrounding and when the fission process operates most effectively and it is also the energy most suitable for neutron radiography. Accelerator and isotopic source neutrons are mostly born at higher energies, up to about 14 MeV, and so the moderation factor (neutrons lost in the energy-reduction process) for these sources is usually poorer than that for a nuclear reactor.

This fission process produces a considerable amount of heat and this is usually removed by a stream of coolant (water or gas) flowing in specially constructed cooling passages in the reactor core. This coolant then loses heat in a conventional heat exchanger.

Control of a nuclear reactor is achieved by simply removing neutrons from the process and thereby stopping the chain reaction. The neutrons are removed by inserting a neutron absorbing material into the reactor core and regulating the extent of the insertion in order to maintain the reactor at a steady operating power.

Thus the nuclear reactor is a device that produces fast and slow neutrons, gamma rays and charged particles, all in prolific quantities and at present it is the most widely used neutron source for neutron radiography.

1.1.3.4 Sub-Critical Assemblies

The output of a non-reactor neutron source can be boosted by incorporating it into a sub-critical assembly. This is a small thermal neutron reactor which has been 'under-designed' so that it is not capable of sustaining a chain reaction. The neutron source is placed in the reactor core and provides the supply of neutrons to keep the reaction going. The fission processes in the fuel can now be regarded as a multiplication phenomena in that the neutron output of the source is enhanced by the reactor neutrons.

Crosby et al [Ref. 17] have investigated such a system for neutron radiography using a Ca^{252} source with a water moderator. This assembly had a multiplication factor of about 30. The factor is dependent upon the value of the effective multiplication constant (K_{eff}) for the system. This constant may be regarded as the ratio of the number of neutrons in one generation, compared to that of the next generation. So if $K_{eff} = 1$ then the neutron population is sustained from one generation to the next. This condition is called critical and is that which applies to the type of thermal neutron reactor described in the previous section. The sub-critical assembly, however, has a K_{eff} of less than one and the neutron density will depend upon the neutron emission from the neutron source. Figure 1.5 shows the relationship between the multiplication factor and K_{eff} and it can be seen that K_{eff} must be greater than about 0,9 before a useful multiplication is obtained. In practice the choice of K_{eff} is made from considerations of safety and cost. A balance must be found whereby the system produces a useful multiplication but K_{eff} is sufficiently far from unity to ensure that the reactor will not go critical. If criticality is possible then stand-by control absorbers must be added to the assembly. Clearly if this situation were possible then it would be more practical to build a critical nuclear reactor in the first place.

So safety requirements will limit the K_{eff} of a sub-critical assembly to about 0,99 – although this will be dependent upon the safety philosophies followed by the local reactor licensing authority – and a corresponding multiplication factor of about 30. This would be a useful increase, but its cost effectiveness must be examined by comparing the cost of the assembly with other types of neutron sources of the same intensity.

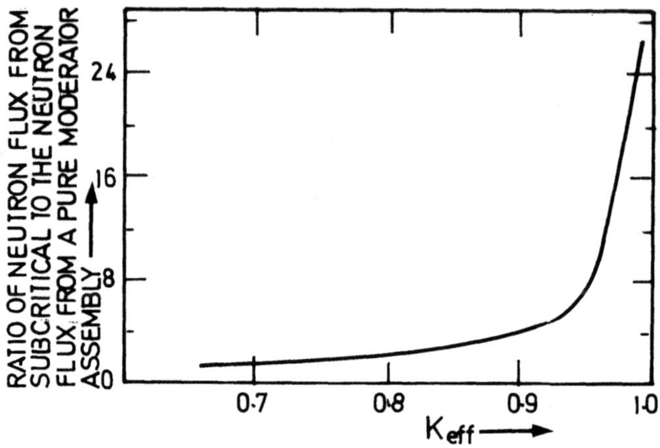

Fig. 1.5 Increase in Neutron Flux as Subcritical Size is increased. (After Bouchey, Int. J. Non-Destr. Test. 2, 1971, p. 350).

1.1.3.5 A Comparison of Neutron Sources

Table 1.3 shows a summary of the properties of some typical facilities for neutron radiography by Hawkesworth and thus provides a convenient format for comparison of the sources, in that it provides a practical criteria for the comparison in the times required to produce a radiograph. It can be seen that the reactor sources are approximately an order of magnitude better than accelerators and ^{252}Cf sources, on this basis. Moreover, other criteria such as cost and mobility may take precedence.

1.1.4. Neutron Beam

So far we have discussed various ways of producing neutrons without considering whether these neutrons will be suitable for neutron radiography. Now, the radiographic process must use a radiation which has a high probability of reacting with the material of the sample, and it is usual to describe this probability of interaction as an effective target size called the 'cross section'.

1.1.4.1 Nuclear Cross Sections

The cross section quantifies the probability that a reaction will take place between the neutron (travelling at some effective velocity) and the target material; it can be considered as a target size and it is measured in cm^2. There are several types of cross section but the two that are of principal interest to neutron radiographers are the absorption cross section and the scattering cross section. The total cross section is the sum of these two.

The unit of cross section is the 'barn', which is 10^{-24} cm^2, and typical cross sections vary from a few millibarns to several thousand barns.

Table 1.3 Summary of the Properties of some typical Facilities for Neutron Radiography [**Ref. 54**].

Source	Collimator position	Collimator entrance flux [c] $(cm^{-2}.s^{-1})$	Typical beam characteristics: Intensity at detector $(cm^{-2}.s^{-1})$	L/D ratio	Cd ratio	Imaging technique [a]: Detector	Film [b]	Neutron exposure time (s)
Multi-purpose research reactor	Radial	10^{14}	10^{8}	250	2 - 5	In	D2	10^{2}
	Tangent.	10^{13}	10^{7}	250	10 - 50	Dy	D2	10^{3}
	Cold source	2×10^{11}	10^{6}	100	∞	Gd	D2	2×10^{3}
Radiography reactor [d]	Radial	10^{12}	10^{6}	250	2 - 5	In	D2	10^{4}
	Tangent.	4×10^{11}	2×10^{6}	100	10 - 50	Dy	D4	10^{3}
Be(d,n); 3 MeV; 400 μA.Accel.	Radial	3×10^{9}	2×10^{5}	33	5 - 20	Dy	D7	5×10^{3}
Be(γ,n); 5.5MeV, 100 μA Accel.	Radial	4×10^{8}	8×10^{4}	18	5 - 20	Dy	D7	5×10^{3}
T(d,n) + U; 120 keV, 7 mA Accel.	Radial	10^{8}	2×10^{4}	18	5 - 20	Dy	D10	8×10^{3}
5 mg ^{252}Cf Isot. + sub-crit. assembly	Radial	3×10^{9}	2×10^{5}	18	2 - 10	Dy	D7	5×10^{3}
1 mg ^{252}Cf Isotope	Radial	2×10^{7}	10^{4}	12	5 - 20	Gd	D10	4×10^{3}

a) The transfer technique has been used as the main example here because of its value to the nuclear industry. The neutron exposure times assume a film exposure time $> 5 \times T_{\frac{1}{2}}$.

b) Films from the Agfa-Gevaert range. Other manufacturers offer a closely parallel series of films.

c) See Section on Collimation.

d) A small reactor designed to make neutron radiography as convenient and economic as possible.

The cross section of the elements and their isotopes vary with the energy of the bombarding neutrons (see Fig. 1.6) in general the lower the energy the higher the cross section. This fact provides the opportunity to increase the transmission of the neutron through the sample by using a neutron energy that coincides with a region of low cross section in the sample (see Neutron Beam Filters).

The transmission of a neutron through a sample may be expressed by considering the rate at which the neutron intensity reduces as it passes through the sample material.

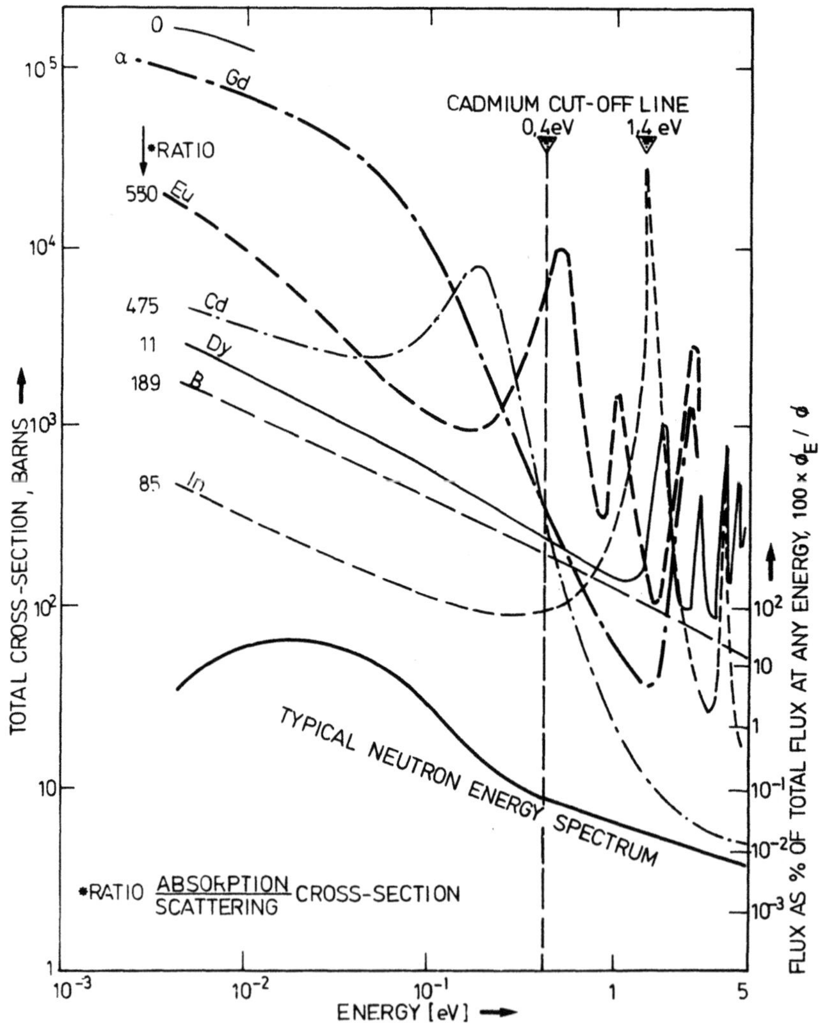

Fig. 1.6 Total Cross-Section Curves for Boron, Cadmium, Indium and Dysprosium.

This is given by

$$-\frac{d\phi}{dx} = \phi\sigma N \qquad (1)$$

where ϕ = neutron intensity, i.e., number of particles passing across unit area in unit time, n cm^{-2} s^{-1}
X = specimen thickness, cm
σ = microscopic cross section, cm^{-2}
N = number of target nuclei per unit volume, cm^{-3} (see Appendix 1.5 for a calculation of N)

Rearranging and integrating gives:

$$\phi = \phi_0 e^{-N\sigma x} \qquad (2)$$

where ϕ = neutrons transmitted through the sample, n. cm^{-2} s^{-1}
ϕ_0 = neutrons incident upon the sample, n. cm^{-2} s^{-1}

The ratio between these two neutron fluxes is called the transmission, i.e. transmission

$$B = \frac{\phi}{\phi_0} = e^{-N\sigma x} \qquad (3a)$$

It should be noted that there are several ways of expressing the cross section of a material, i.e.:

a) microscopic cross section, cm^2
b) macroscopic cross section, cm^{-1}
c) mass absorption coefficient, cm^2 g^{-1}

The first is the basic unit and, as stated earlier, is measured in barns. The second is the product $N\sigma$; this is given the symbol Σ and it is the total target area for a given neutron interaction presented by a cubic centimeter of material.

Thus, for this case equation (3) can be rewritten:

$$B = e^{-\Sigma x} \qquad (4)$$

and it can be seen that the use of the macroscopic cross section Σ simplifies the use of this equation.
The third form, the mass absorption coefficient, is denoted by the symbol μ_m and is related to the macroscopic cross section by the relationship

$$\mu m\,(cm^2\,g^{-1}) = \frac{\Sigma\,(cm^{-1})}{\rho\ \ (g\,cm^{-3})}$$

where ρ = density

Thus the mass absorption coefficient is the total target area for a given neutron interaction per cubic centimeter of material per unit density. It therefore more conveniently expresses values for solids and gasses whose densities normally differ by several orders of magnitude.

In this case equation (3) is expressed as

$$B = e^{-(\frac{\Sigma}{\rho} \cdot x \rho)} = e^{-\mu_m \cdot x \cdot \rho} \qquad (3b)$$

where the unit $x\rho$ has the dimension of g cm^{-2}.

Values of Σ and μ_m are given in Appendix 1.2 and an example of the calculation of Σ for a compound is given in Appendix 1.5.

It should be noted that there is also a cross section called the 'linear absorption coefficient' with the dimension cm^{-1}. This is the same as the macroscopic cross section but is more usually used in describing alpha, beta and gamma interactions.

1.1.4.2 Moderation

The neutrons from all available sources are born with high energies, up to about 14 MeV, whereas neutron radiography requires neutrons having energies of about 0.03 - 10 eV. Thus they must be slowed down to the thermal/epithermal neutron range, and so here we have the same problem that exists in the nuclear reactor, that of moderation, and of course the same solution can be used. A moderating material, usually water or beryllium, is placed around the radioisotope neutron source or the target of the accelerator, to produce a neutron energy spectrum similar to that of the nuclear reactor. The inherent advantage of the nuclear reactor now becomes clear; it already has a predominantly low-energy spectrum, and furthermore reactor neutrons are born at a lower energy (approx. 2 MeV as compared to about 14 MeV for an accelerator) and so fewer neutrons are lost in the moderation process.

A prime requirement for a moderator is that it shall slow down neutrons without absorbing them, and so good moderating materials will have a large scattering cross-section and a small absorption cross-section. Each time a neutron interacts with the moderator there is a probability that it may be absorbed and so the fewer the number of collisions required to produce a thermal neutron the better for moderation. Once it has reached thermal energy the neutron will continue to bounce around until it is finally captured.
In order to reduce the energy of a neutron in the fewest number of collisions the amount of energy lost per collision must be as large as possible. It can be shown that the log of the average energy loss is, to a close approximation, inversely proportional to the atomic number, and so for the best 'energy loss' conditions light materials should be used. Hence moderators are those materials that have low atomic number, low absorption cross-section, and a high scattering cross-section and typical moderator materials are light water, heavy water, and carbon.

1.1.4.3 Collimation

Having produced a source of low energy neutrons we now have to form them into a usable beam. The neutrons will move about the moderator in a completely randow manner, and unlike electrons cannot be focused. The best that can be achieved is to contain the neutron source and the moderator within a neutron-absorbing shield and allow some of the neutrons to stream down a hole in this shield.
To achieve this containment the walls of the collimator are lined with a neutron opaque material which will prevent stray neutrons entering the system via the collimator walls and will also reduce low angle scattering within the system. This lining material must have a high cross section to neutrons, and the secondary radiation produced by neutron absorption with this lining must have a low probability of being recorded. These requirements lead to the use of boron, cadmium*), dysprosium, europium, gadolinium and indium as lining materials.
The angular spread of the emerging beam will be confined by the length-to-diameter ratio of the collimator hole, and so to ensure a narrow beam-spread a collimator usually has a high L/D ratio. Near parallel neutron beams are achieved by using a bundle of small tubes or a stack of equispaced plates within the collimator hole. With these arrangements the L/D ratio of the collimator is that of an individual tube or the gap between the plates. This ratio can be made very large, but this type of collimator suffers a considerable loss in beam intensity and produces a pattern of circles or lines on the radiographic image. An alternative method, widely used for neutron radiography, is the divergent collimator. This has a relatively small inlet aperture, and the collimator hole diverges uniformally along its length. With this arrangement the angular spread of the neutron beam reaching the object only depends on the source size and the distance. These are shown in Figs. 1.7A and 1.7B **)
An estimate of the flux emerging from a collimator tube [Ref. 11] is given by:

$$\bar{\phi} = \frac{A}{16}\left(\frac{D}{L}\right)^2 \left(\phi_i + \frac{1}{\Sigma}\frac{\partial \phi}{\partial z}\right) \qquad (5)$$

where ϕ_i = neutron intensity at entrance to the collimator, n cm^{-2}s^{-1}
$\bar{\phi}$ = neutron emission at the exit from the collimator, n s^{-1}
A = collimator area at exit, cm^2
D = diameter of inlet aperture, cm
L = length of collimator, cm
Σ = macroscopic total cross-section of moderator, cm^{-1}
$\partial\phi/\partial z$ = flux gradient at inner face of collimator

As the flux gradient is usually small this can be further approximated by

$$\bar{\phi}. = \frac{A}{16}\left(\frac{D}{L}\right)^2 \phi_i \qquad (6)$$

---

*) The use of cadmium for the direct method is limited by the high energy gamma emitted after neutron capture.

**) In the diagrams in Fig. 1.7 the lengths L and L_s plus L_f are the radiographic lengths of the collimator, and these are assumed to be approximately equal to the physical length of the collimator.

and so

$$\phi_0 = \frac{\overline{\phi}}{A} = \frac{1}{16}\left(\frac{D}{L}\right)^2 \phi_i \qquad (7)$$

where ϕ_0 = neutron flux at outer face of collimator, n $cm^{-2}s^{-1}$.

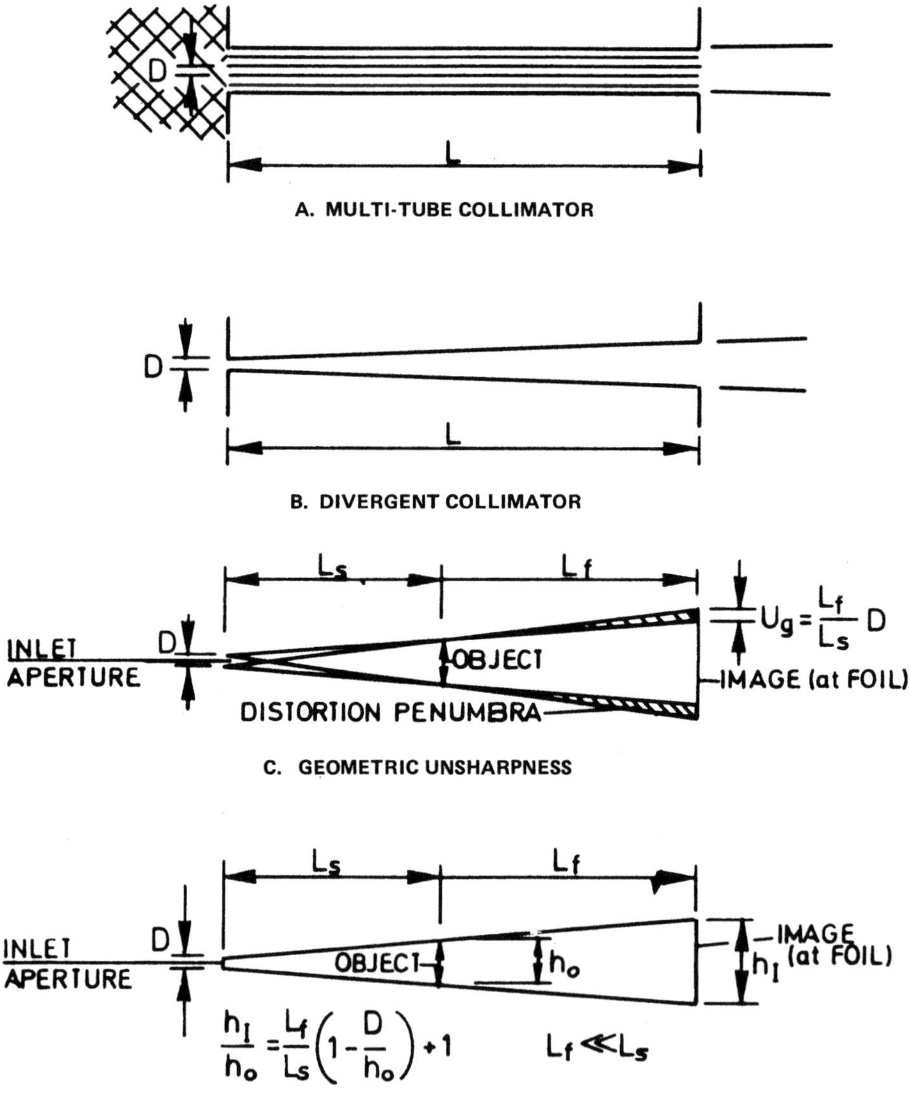

Fig. 1.7 Neutron-Beam and Collimator Geometry.

Now the fraction of neutrons lost due to collimation will be the ratio:

$$\frac{\text{flux at the entrance}}{\text{flux at the exit}} = \frac{\phi_i}{\phi_o} = 16 \left(\frac{L}{D}\right)^2 \qquad (8)$$

The collimator length to inlet diameter ratio (L/D) is called the collimator ratio. Expression (8) assumes that all of the neutrons in the collimator originate at the entrance aperture, but in practice some will come from the walls of the collimator [Refs. 12,22] adjacent to the aperture. Hawkesworth [Ref. 12] has shown that the total flux is given by

$$\phi_o = \frac{\phi_i}{16} \left(\frac{D}{L}\right)^2 \left(1 + \frac{2l}{L}\right) \qquad (9)$$

where l = length of collimator wall which emits neutrons.
The lenght l varies with different types of neutron radiography unit in that it is usually a section of the collimator that is not lined.
A lining close to the source will depress the flux and so with low intensity sources it is necessary to leave an unlined section at the beginning of the collimator. High flux reactor sources are not so concerned with this flux depression and in most designs the collimator is lined along its full length. For sources of low intensity the unlined length is usually about two diameters (D) long and so equation (9) can be modified to

$$\phi_o = \frac{\phi_i}{16} \left[\left(\frac{D}{L}\right)^2 + 4\left(\frac{D}{L}\right)^3\right] \qquad (10)$$

for simplicity this can be expressed in the same form as equation (8) by a close approximation, i.e.

$$\frac{\phi_i}{\phi_o} = F_c \left(\frac{L}{D}\right)^2 \qquad (11)$$

where the value of F_c is close to 12.
Hawkesworth found reasonable agreement with equation (9) for small L/D ratios (<30) on accelerator units. Matfield examined published data relating to a large number of neutron radiography units of all types and found that the constant F varied from about 1 to 100 although this may well be due to uncertainties in the value of the source flux, as this is often difficult to measure accurately, and also the uncertainties in the assumption that the collimator aperture is the true neutron aperture.

The length to diameter (L/D) ratio for a collimator effects both the resolution and the collimator efficiency and so it is widely used as a simple means of characterising a collimator. The resolution of the collimator can be described by considering the effect of the width of the radiographic dimensions of the collimator on the unsharpness of the image,(as shown at Fig. 1.7c), where the dimension L_f is shown grossly oversize in order to ensure clarity of the diagram (also see footnote to page 15).

This unsharpness is expressed by $U_g = \frac{D}{L_s} L_f$ (12)

where U_g = geometric image unsharpness, cm
D = source-aperture size, cm
L_f = image to object distance, cm
L_s = source to object distance, cm.

Usually $L_f << L_s$ and so the geometric unsharpness is linearly dependent upon the inverse of the collimator ratio.

To minimise unsharpness the aperture size D and the distance L_f must be small and L_s must be large. These requirements lead to a direct conflict with the desire to achieve maximum intensity, for this requires D to be large and L_s to be small. So we must compromise, and the practical design is based on a judgement of the unsharpness that can be tolerated, in conjunction with a workable neutron flux.

It is clear from the above discussion that the length to diameter ratio of a collimator is an important characteristic, but we can achieve a particular L/D ratio by using an infinite variation of length or diameter, so is it better to have a long and wide collimator or a short and narrow one? The effect of collimator size has been measured by Barton [Ref. 13] and others and it is clear that the answer cannot be given without a number of qualifications, for we are concerned with the interdependence of resolution, contrast, intensity, neutron/gamma ratio and the attenuation produced by the specimen. Nevertheless the following general statements can be made:

(a) the contrast will vary with the intensity of the gamma rays in the beam
(b) for any particular sample the neutron attenuation will be different to the gamma attenuation.

Due to (a) the transfer method (see 1.1.6.2) usually produces better contrast than the direct method, and due to (b) the presence of gamma rays in the beam will reduce the contrast for the direct method unless the gamma attenuation of the sample is high, say > 80%.

(c) The neutron flux increases more rapidly than the gamma flux as the aperture size increases,
(d) the neutron flux and the gamma flux decrease at about the same rate as the collimator length increases,
(e) the slow neutron intensity decreases more rapidly than the fast neutron intensity as the length of the collimator increases.

Due to (c) and (d) long and wide collimators have better neutron to gamma ratios and hence better contrast for the direct method. But due to (e) this will produce a lower slow-to-fast neutron ratio. This will reduce the contrast for a sample with high scattering cross-section, but it will increase the penetration of a sample with a high absorption cross section.

(f) the resolution improves with an increasing L/D ratio.

To fully achieve (f) there must be no loss of contrast and so if the available collimator is short and narrow with a poor neutron to gamma ratio the best resolution may be achieved by the direct method. Similarly the resolution may be effected when a sample with a large scattering cross section is radiographed in a beam with a high fast neutron component.

(g) the larger the inlet aperture the greater is the displacement of the moderator,
(h) the longer the collimator the greater is the attenuation loss due to the collimator atmosphere.

Statements (g) and (h) are mainly relevant to low intensity sources and for this type of source (g) often outweighs all others for it has a direct effect on the neutron intensity at the entrance to the collimator. Furthermore the spatial neutron flux in a small source moderator can vary across the collimator aperture and thus for this type of source the collimator is usually made short and narrow. For a collimator filled with air the loss is about 5% per metre, and where helium is used this is reduced to $<1\%$ per metre. Hence the neutron attenuation due to the collimator atmosphere is only significant for air and once again leads to the use of short collimators for low intensity sources.

When we collect these statements together it is clear that the original question on the shape of the collimator must include some information on the type of sample to be radiographed, the acceptable results, etc. In practice the construction of the neutron source usually sets the overall limits to the width and length of the collimator and where possible the best solution is to make the aperture as wide as possible and stop it down with a range of insertable aperture plates.

1.1.5 Neutrons Applied to Radiography

So far we have followed the progress of a neutron, born with high energy, progressively losing this energy by successive collisions within a moderating material, until it finally escapes along a collimator tube. At the outer end of this tube there is a sample to be radiographed, and this is duly struck by the neutron. What happens now is well described by a comparison with X-rays. This comparison is shown in Figure 1.8 where the mass attenuation coefficients of X-rays and thermal neutrons have been plotted against the atomic numbers of most of the elements. X-rays show a continuous curve and so any two materials having a similar atomic number will lie close to each other on the curve and consequently have similar mass absorption coefficients. Both materials will, therefore, attenuate an X-ray beam by about the same amount and so it will be difficult for a detector to discriminate one from the other. The attenuation of neutrons however, is a function of the nucleus rather than the density of electrons in a material, and it is frequently found that adjacent-number elements, for example boron and carbon, show marked differences in neutron-attenuation coefficient and are therefore readily discriminated. Hydrogen has a high neutron attenuation coefficient, and so it is possible to detect rubbers and plastics. Conversely dense materials such as lead and tungsten have low coefficients and are readily penetrated by thermal neutrons.

Thus we find that the two radiographic processes, X-ray and neutron, are often complementary. X-rays are stopped by dense materials and pass through light ones, and in many instances neutrons have the reverse qualities. Neutrons will penetrate the body of a large metal valve to give a good image of an internal rubber seal. For the X-rays to record this seal would require a long exposure which would probably obliterate most of the other detail on the radiograph. However, if the valve were inside a thick polythene case then the X-rays would penetrate this with negligible attenuation, whereas the neutrons would have difficulty in producing a radiograph.

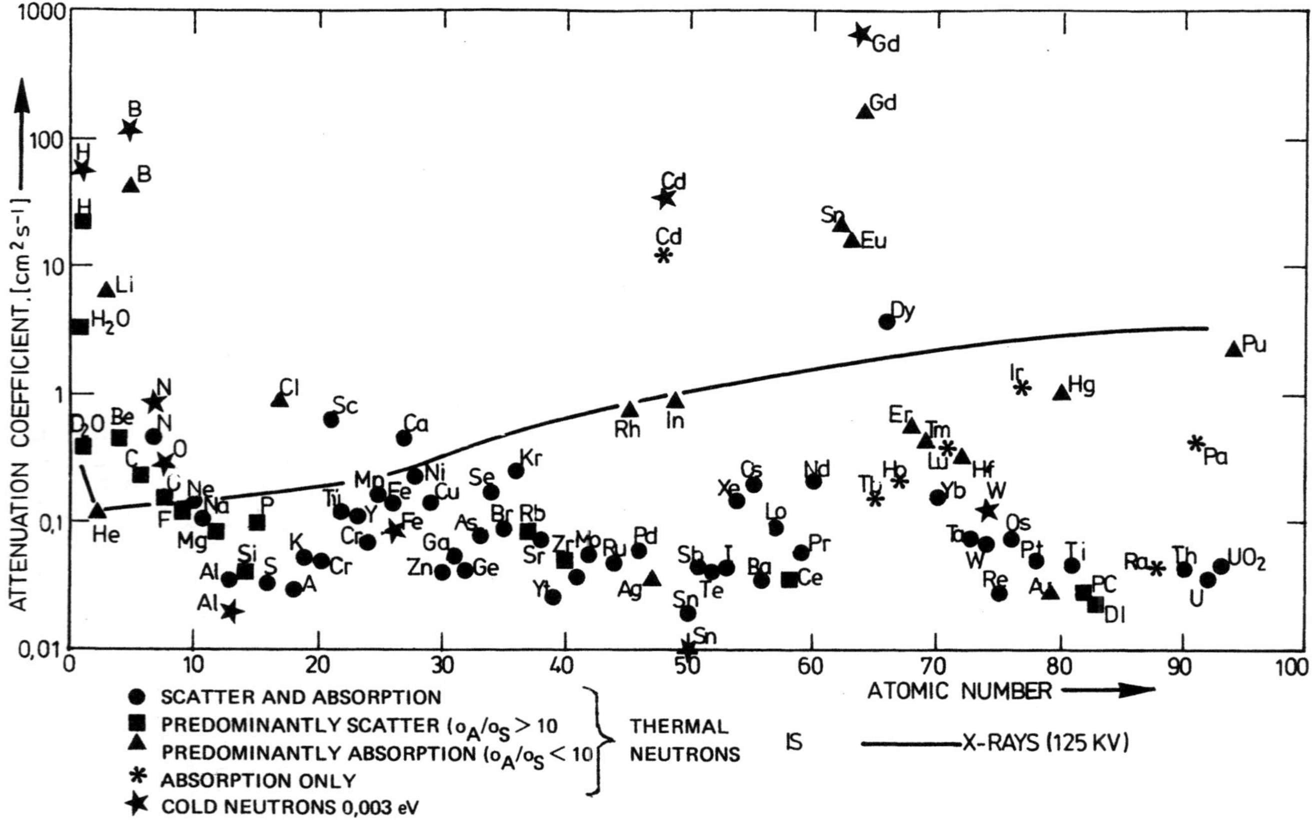

Fig. 1.8 Neutron and X-Ray Mass Attenuation Coefficients for the Elements.

1.1.6 Neutron Image Detectors

Unlike X-rays, neutrons have very little direct effect on photographic film, the quantity of silver in the grains of a photographic emulsion being insufficient to produce much neutron absorption. The thermal neutron half-life is relatively long, and it would take many hours to obtain a usable image. Thus for neutron radiography it is necessary to use a slightly different technique. The method normally employed uses an intermediate foil which converts the neutron image into alpha, beta, or gamma radiation, and it is this secondary emission which is detected by a photographic film.

1.1.6.1 Direct Technique

With this method an atom in the foil absorbs a neutron and it promptly emits other actinic radiation. This is called the direct technique (see Figure 1.9), because the foil is placed directly into the neutron beam in contact with the photographic film. When a metal foil, such as gadolinium, is used the induced radiation is an electron. Alternatively a scintillator screen containing a mixture of lithium-6 and zinc sulphide can be used. On absorbing a neutron a lithium atom emits an alpha particle and this then strikes the zinc sulphide, which in turn emits a light photon. As the above processes are continuous reactions this type of foil and scintillator screen can be used with low neutron fluxes and long integrating exposures, and because the film is in contact with the converter during the neutron exposure, all of the forward emitted radiation takes part in the exposure of the film. Thus the direct technique is fast, the scintillator screen type of converter being 30-100 times faster than metal foils. The particular advantage of the gadolinium metal foil is that the neutrons are absorbed in a very thin layer of the foil and the emitted electrons have a short range, and so a resolution of about 12 μm can be achieved.
For nuclear applications the direct technique has a disadvantage. If the object is radioactive it will invariably emit gamma rays and these will mingle with the neutron beam and produce a second, auto-radiographic image on the film. As these gamma rays come from a different source from that of the neutrons the images will be different and the film will be 'gamma-fogged'. Fortunately this can be avoided by using the transfer technique or track-etch recorders (see 1.1.7.2)

1.1.6.2 Transfer Technique

This method (see Fig. 1.9) relies on the build-up of radioactivity in the foil produced by neutron absorption. In this way an activation image is formed in the foil, and this is subsequently transferred to a photographic film in a dark-room by placing the foil and film in contact and allowing the decay radiation from the foil to produce the latent image in the film.
With this method the decay process starts during neutron exposure in the beam and so some of the emitted radiation is lost. This makes the transfer technique slower than the direct method but this disadvantage is more than compensated for by the fact that since the foil is insensitive to gamma rays the method can be used in gamma-ray fields of any intensity.

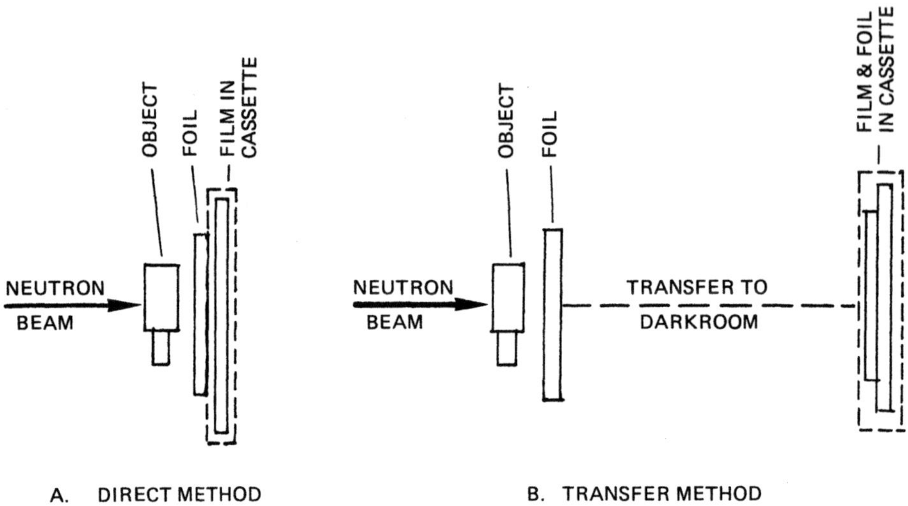

Fig. 1.9 Direct and Transfer Method for producing a Neutron Image.

With the transfer method only the foil is present in the beam, and the activity builds up exponentially according to

$$S = \phi \sigma N (1 - e^{-\lambda T}) \qquad (13)$$

where S = activity, disintegrations s^{-1}
ϕ = neutron flux, n $cm^{-2}s^{-1}$
σ = microscopic neutron cross-section, cm^2
N = number of atoms in the sample
λ = $0.69/\tau$
τ = half life, s
T = irradiation time, s

It can be seen from expression (13) that the activity will gradually reach a maximum, depending upon the neutron flux and the cross-section. The foil activity will eventually saturate, and since neutron fluxes below about 10^4 n $cm^{-2}s^{-1}$ will not provide sufficient intensity to store enough energy in the foil to produce an acceptable image on a photographic film, this restricts the use of the transfer technique to moderate and high intensity sources.
The correct exposure for the transfer method is determined by both the length of time that the foil is in the neutron beam and the time that the image is allowed to transfer to the film (see Fig. 1.10). Both of these may be varied, as any practical product of the irradiation fraction (activity given to foil) and transfer fraction (part of this activity transfered to the film) will give the same result. Table 1.4 gives the characteristics of some converter foil materials.

Table 1.4 The Characteristics of Some Possible Neutron Radiography Converter Materials [Ref. 14]

	Material	Abundance of Parent Isotope, %	Mode of Production of Active Isotope	Cross-Section barns	Half-life	Emission Type	Emission Max. Energy, MeV
D	Lithium	7.4	$Li^{6}(n,\alpha)H^{3}$	935	STABLE	α	4.7
D	Boron	19.5	$B^{10}(n,\alpha)Li^{7}$	3,837	STABLE	α	2.3
D	Rhodium	100	$Rh^{103}(n,\gamma)Rh^{104}$	144	43 s	β	2.41
			$Rh^{103}(n,n^{1})Rh^{103m}$		57 min	X-ray	0.02
			$Rh^{103}(n,\gamma)Rh^{104m}$	11	4.4 min	β	0.5
D	Silver	51.4	$Ag^{107}(n,\gamma)Ag^{108}$	44	2.4 min	β	1.64
						β	0.43
		48.7	$Ag^{109}(n,\gamma)Ag^{110}$	110	24.5 s	β	2.87
			$Ag^{109}(n,\gamma)Ag^{110m}$	3	254 d	β	1.5
							0.66
D	Cadmium	12.3	$Cd^{113}(n,\gamma)Cd^{114}$	20,000	STABLE	γ	9
T	Indium	95.7	$In^{115}(n,\gamma)In^{116}$	45	14 s	β	3.3.
							0.44
			$In^{115}(n,\gamma)In^{116m}$	154	54 min	β	1.0
							0.42
D	Samarium	13.9	$Sm^{149}(n,\gamma)Sm^{150}$	41,500	STABLE	γ	
		26.6	$Sm^{152}(n,\gamma)Sm^{153}$	210	46.7 h	β	0.8
							0.1
D	Gadolinium	14.7	$Gd^{155}(n,e)Gd^{156}$	58,000	STABLE	e	0.14
		15.7	$Gd^{157}(n,e)Gd^{158}$	240,000	STABLE	e	0.13
T	Dysprosium	28.1	$Dy^{164}(n,\gamma)Dy^{165}$	800	2.3 h	β	1.29
							0.095
			$Dy^{164}(n,\gamma)Dy^{165m}$	2,000	1.26 min	β	1.04
							1.108
T	Gold	100	$Au^{197}(n,\gamma)Au^{198}$	98.8	2.69 d	β	0.962
							0.412

D = direct method T = transfer method

1.1.6.3 Dynamic Imaging Methods

The exposure times for the methods described above, even assuming that intense neutron beams from nuclear reactors are used, are usually greater than one second so they are essentially 'still' techniques. To produce high-speed, or flash, radiography requires a neutron source that will produce an ultra-high intensity flash of neutrons lasting a few milliseconds, such as can be achieved with a pulsed nuclear reactor. This is a reactor which is made sub-critical by removing part of the uranium fuel in the reactor core. To produce a pulse the missing fuel rod is passed rapidly through the core, causing the reactor to go critical for a very short time and producing a large pulse of neutrons. Pulse widths of a few milliseconds are possible, so rapid motion may be arrested in

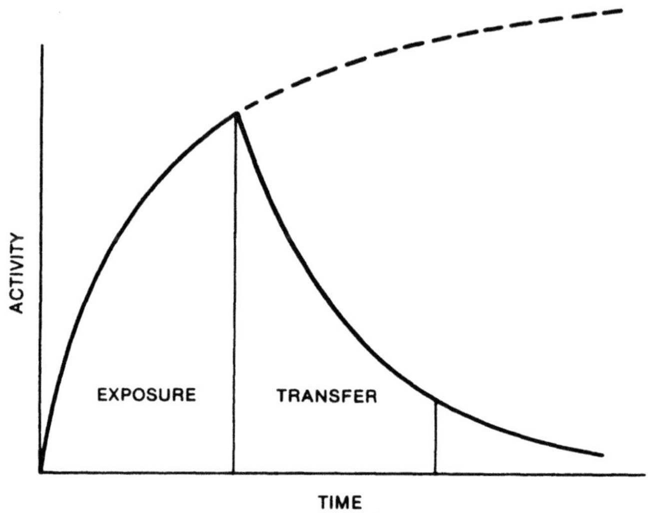

Fig. 1.10 Irradiation/Transfer Curve.

the radiograph. A limitation of this system is that most pulse reactors are only capable of a few pulses per day. Examples of this technique have been reported by Mullender and Hart [Ref. 34].

A method of producing a moving picture is to observe the light output from a scintillator screen with a television system (see Figure 1.11). With most neutron radiography units the neutron beam strength is too low to give sufficient light intensity to be seen on a TV monitor, so an image intensifier is placed between the scintillator screen and the camera. This boosts the light from the screen by about 10^4 x , and object movement at up to about 3 m/s can be observed with good definition. Where a neutron beam strength of about 5×10^8 n $cm^{-2}s^{-1}$ or greater is available the image intensifier might be omitted, provided the spectral output of the screen is properly matched to the spectral response of the TV tube.

A more sophisticate system [Ref. 35] uses an intensifier tube which has a front gadolinium screen, the secondary electrons from which are accelerated on to a scintillator screen (this assembly is often called a neutron camera). The tube is [Ref. 35] claimed to have a resolution of 30 line pair per cm and gains of 3-5×10^6 for cold and thermal neutrons. The tube can be used with a T.V. camera/monitor to provide a remote display and hard copies of the images can be provided.

1.1.7 Image Recorders

At present the recording materials in general use for neutron radiography are photographic film and track etch films, each of which is described below.

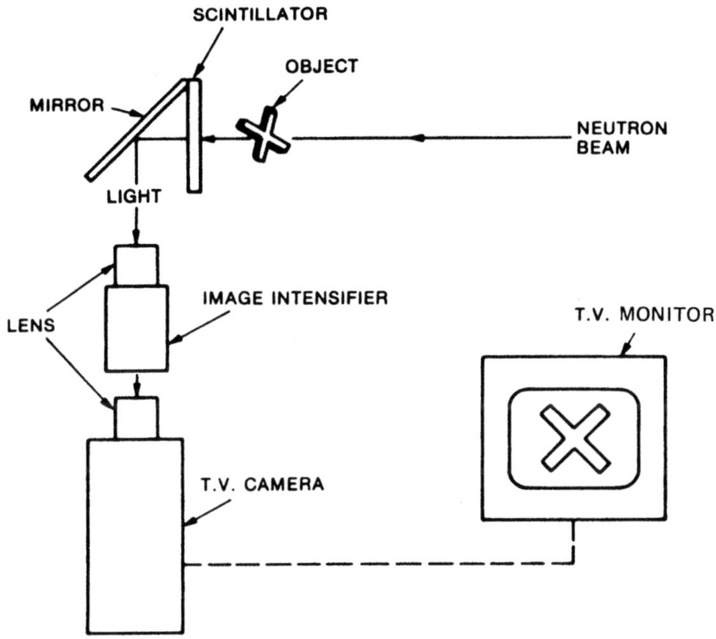

Fig. 1.11 Television System for Neutron Radiography.

1.1.7.1 Photographic Film and its Characteristics

No special films are available for neutron radiography, and standard X-ray and photographic films are normally used. These consists of a base material with a gelatine coating in which are dispersed minute grains of silver halides, the grain size varying from 0.1 to 3 μm, depending on the type of film.

When photons or electrons fall onto such an emulsion electrons and positive silver ions migrate to points of imperfection in the silver-halide crystals and on arrival some of the silver ions are reduced to metallic silver to form the latent image. When developed with suitable agents the silver halide at the latent image is further reduced to metallic silver, and the unaffected halide grains are subsequently dissolved away by the fixing solution, leaving a black metallic-silver image. The latent image is distributed throughout the entire emulsion, and the density of activated silver grains increases almost uniformly with exposure. Development, on the other hand, proceeds from the surface in a complex manner, grains at different depths converting at different rates as the developer penetrates the emulsion. After development the silver grains are viewed as a two-dimensional array, and the distribution in depth is not apparent. This distribution causes the grains to appear as groups, and contributes to the characteristic graininess seen in photographic films.

The speed of a film is essentially a measure of the blackening produced by a given exposure. Blackening is better produced by largegrain emulsions, where fewer developed grains per unit area are needed to give a recognisable change. But the larger the grain the poorer the resolution, so whilst large-grain emulsions, such as X-ray films, will give rapid and acceptable results for normal inspection purposes, for high resolution work fine-grain film should generally be used.

Contrast and resolution are dependent qualities, and in order to obtain good resolution on a film the contrast, i.e. the density variations, must be adequate. This is well demonstrated by exposing and developing a double-emulsion X-ray film and then scraping away part of one emulsion. It is possible for the image on the double emulsion to appear to be of higher resolution then that of the single emulsion owing to the greater contrast. So although the use of thin emulsions will lead to better resolution, sufficient image density must be generated to achieve the best contrast for the type of film being used.

The relationship between exposure and film density is expressed by a characteristic curve, where the density is plotted against the logarithm of the relative exposure (see Fig. 1.12) and density as

$$D_e = \log_{10}\frac{I_o}{I} \qquad (14)$$

where I_0 = intensity of incidence light on the film
I = intensity of transmitter light from the film.

The neutron radiographic technique produces an activation image of the

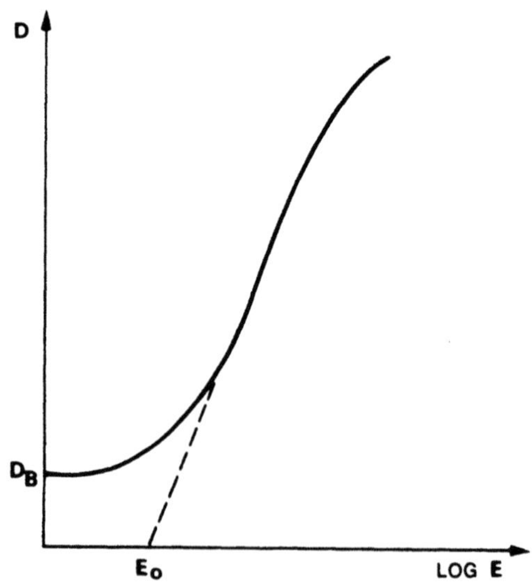

Fig. 1.12 Typical Characteristic Curve for a Radiographic Film.

specimen and its background on a foil, and the radiation from these regions will determine the relative exposure of the adjacent areas of the film. The corresponding density difference will depend on the part of the characteristic curve upon which these exposures fall. The steeper the slope of the curve in this region, the greater will be the density difference and hence the greater the contrast. As high contrast is a basic requirement for a good radiograph every endeavour is made to work on this steep part of the curve.
A comparative list of some of the photographic films that have been used for neutron radiography is given in Table **1.5**.
Photographic films suffer from dimensional instability due to changes in humidity, temperature, and processing [Ref. 49]; an average variation could be 0,002 in./in. for triacetate base film and 0,001 in./in. for polyester films. However most modern films are manufactured with polyester bases.

1.1.7.2 Track-etch Recorders

Photographic film has some disadvantages. It must be developed under darkroom conditions, and the dimensions of the film can change with variations in temperature, humidity, and development. Alternative materials are feasible, and experiments with detectors for health physics applications in the early 1960s showed that the damage tracks caused by bombarding mica with heavy particles could be 'fixed' by etching in an acid solution and that these could be observed with an electron microscope. The technique was developed for direct viewing by using a combination of a boron (or lithium) foil with a nitrocellulose film, a combination which utilises the neutron/alpha-particle reaction in boron, the alphas causing surface-damage pits in the cellulose and, unlike the beta particles emitted from the metal foils currently in use with the transfer process, the alpha particles take short and relatively straight paths through a material and so give good resolution.

In 1973 Kodak, France, marketed a nitrocellulose film for neutron radiography coated on both sides with lithium borate. After irradiation the lithium borate is removed by washing and then the film is etched. The contrast is very low but this can be significantly improved by printing on to a copying film, using a point source enlarger.

Farny [Ref. 36] has made the following comparison between nitrocellulose and photographic film/foil techniques, based upon the use of type CA 80-15 B and a gadolinium foil and type R film (single coated X-ray film):

Direct method

(a) The performance differs little for a given fluence, but as cellulose nitrate is insensitive to γ's a bismuth filter is unnecessary, which leads to a flux gain for the same source and consequently higher overall efficiency.

(b) Using cellulose nitrate the definition is better if the examination can be made on the negative, it becomes comparable if copying is unavoidable due to lack of contrast.

Table 1.5 Approximate Comparison of European and U.S.A. Films [Ref. 32]

Type	Kodak Ltd. (U.K.)	Ilford Ltd. (U.K.)	Kodak Pathé (France)	Gevaert Agfa N.V. (Belgium)	Eastman Kodak Ltd. (U.S.A.	Du Pont (U.S.A.)	Ansco (U.S.A.)	Ref. Speed
Medical and Industrial X-ray	Royal Blue	Gold Seal		Curix RP 1	Royal Blue Type F			
	Blue Brand	Red Seal		Curix RP 2	Blue Band	Type 508		
	Industrex S	Industrial A		Structurix-S		Type 504		
	Standard	Standard						
	Kodirex- Auto Process	Ilfex		Osray T4	Non-screen Medical			
		Industrial G						
	Kodirex				Type F			0.2
			Kodirex	Structurix-D10			Type C	0.3
	Industrex D	Industrial B						0.5
			Analyse					0.6
	Crystallex	Industrial CX		Structurix-D7	Type AA	Type 504	Type A	1
			Definex					1.2
						Type 506		1.4
		Industrial C						1.5
					Type T		Type B	2
		Industrial F						3.1
				Structurix-D4	Type M	Type 510		3.8
	Microtex							4
			Type M					5
					Type R			8
				Structurix-D2				15
Photographic	Royal-X pan	H.P.S. Aerial N		Isopan record	Royal-X Pan			
	Panchro-Royal	H.P.3 sheet		Isopan 27				
	Tri-X pan	H.P.3 roll		Isopan ultra	Tri-X pan			
	Super-XX	Selochrome Pan		Isopan 24	Super-XX			
	Plus-X pan	F.P. 3						
	Super XX Ariel							
		HR Aerial						
	Panatomic X (55)							
Copying		R5.50						

Note: Most of the above list is arranged in approximate order of speed (fast at the top). The precise relative speed data in the right-hand column was supplied by H.P. Leaflets

(c) The contrast is weaker, but can be improved by copying. Moreover, the feasibility of stopping development at intermediate stages without fixing permits the production of many prints of differing contrast from a single exposure.

(d) As in X-radiography, cellulose nitrate film can take the shape of the object during irradiation, or be folded to reach into a cavity.

(e) The customary use of the film in industry usually involves routine printing from the cellulose nitrate, which is a disadvantage considering its simplicity when used in the direct method.

Indirect method (a) The irradiation must be 3-5 times higher to obtain an image comparable to that with Dy and Type R film exposed to infinite

(b) The definition of cellulose nitrate is markedly better (range of α's less than range of $\simeq$ 1 MeV β's).

(c) There is no handling of an active converter.

(d) This film lends itself much better to measurement with a profile projector.

(e) Contrasts are weaker, but this is not always a drawback. For example, when neutron radiographing irradiated fuels, cellulose nitrate is preferable for the examination of fuel-cladding contact, as the difference in contrast between the two materials is very much less.

(f) Because of the linear flux response, there is no saturation of the converter, which can be advantageous when using low-intensity sources. Moreover, the absence of decay and the reduction of exposure times considerably simplifies the whole process of neutron radiography.

(g) Nitrocellulose is a 1/v detector and so it cannot replace indium for epithermal flux work.

Barbalat [Ref. 37] has investigated the effectiveness of several converters when used with Kodak CN 85 nitrocellulose film and found that the relative speeds (in decreasing order) of these were:

a) ^{10}B (75μ)
b) ^{6}LiF (50μ) / ^{11}B (50μ)
c) ^{7}LiF (50μ) / ^{7}LiF (35μ)
d) ^{7}LiF (30μ)

The developer used was 150 g/l of KOH at 40 °C and the developing time was 30 mins. The etching bath was stirred before use and long etching times were avoided in order to prevent sediment formation in the bath due to the camfer removed from the nitrocellulose. Barbalat reported that agitation during the etching causes cloudiness on the film, but not all workers have found the phenomenon. Close temperature control is important so the bath should be placed in a temperature controlled medium.

The CN85 nitrocellulose can be directly examined by placing it between two polarised filters [Ref. 46], or by enlargement with an optical projector. Measurements may also be made on a microdensitometer using the polaroid filters. However the filter does not give any improvement over the direct examination method.

1.1.8 Film and Foil Relationships

For a particular neutron radiographic facility the choise of film and foil will determine the information that can ben recorded in the image. Because the radiographer normally has to make a trade-off between speed and resolution the optimal film and foil combination will change with every type of object to be radiographed. The following sections discuss some of the factors relating to this trade-off.

1.1.8.1 Film and Foil Speed

The speed of films and foils used in neutron radiography have been measured by Berger [Ref. 4], Hawkesworth [Ref. 20] and many other workers, but the results are always relative to the neutron beam which was used. The variation in published values are not large but for the praticing radiographer it is important to have a reliable set of film characteristic curves and one of his first tasks is to carry out a series of calibration measurements on a range of films using different types and thicknesses of foil and screens.
A typical group of these curves for a number of films and foils is shown on Figure 1.11. The data used in this figure were obtained from the DIDO radiography unit at Harwell.

I.1.8.2 Film and Foil Resolution

Methods of recording images in neutron radiography depend on an intermediate detector foil which emits radiation to which the film is sensitive. This introduces unsharpness into the recording process, for while the neutrons passing through the specimen and arriving at the foil may have a narrow angular spread, the radiation emitted from the foil has not, and may enter the film obliquely and be recorded at some small lateral distance from its point of origin in the foil. The magnitude of this effect depends on the range of the particle and the thickness of the detecting foil and film, and is minimised by making these as thin as possible. Reducing the thickness of the foil and film however leads to a reduction in sensitivity of the system to neutrons.
The dependence of film blackening on particle energy is an inverse function of the average energy of the particle, the most efficient energy for this process imately 100 keV [Ref. 27]. This is demonstrated by the superior resolution of a gadolinium foil, which emits an internal-conversion electron of 70 keV, compared to indium and dysprosium which both emit particles of about 1 MeV.

The literature often describes methods of measuring the resolution capabilities of a film/foil combination by radiographing an indicator on which there are a number of objects of graduated size, such as wires or a number of webs between a row of closely spaced holes, and the smallest size that can be discerned by direct viewing is taken as the resolution capability.
Such measurements are not a determination of the resolution of the film/foil combination alone for they include the unsharpnesses in the total radiographic system. Also when the total system unsharpness is greater than the size of the wire of web (which can occur with the smallest sizes) there is an additional complication due to the resulting overlapping of the edges of the image, causing the width of the image to increase and the image contrast to be reduced.
The method usually used is to place a thin opaque knife edge in contact with the foil and assume that because it is in intimate contact with the foil, the geometric unsharpness will be negligible. The measurement will, of course, include unsharpness due to film/foil contact, but if a vacuum cassette is used then this unsharpness can also be assumed to be negligible and the measurement taken as the film/foil unsharpness.
The unsharpness measurement is made by using a microdensitometer to record the density variations across the image of the edge. The method evolved by Klasens [Ref. 48] is usually used to determine the unsharpness from the microdensitometer curve, in which a straight line is drawn to cut the 'S'-shaped density curve at 0.16 x (density range of test object). The projection of this line on the maximum and minimum plateau lines of the density curve gives the unsharpness value (see Figure 1.13).

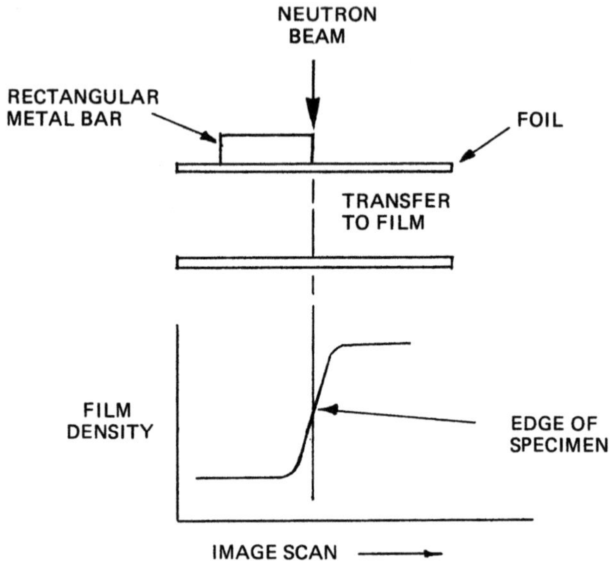

A. EXPOSURE DISTRIBUTION ACROSS THE EDGE OF A SPECIMEN

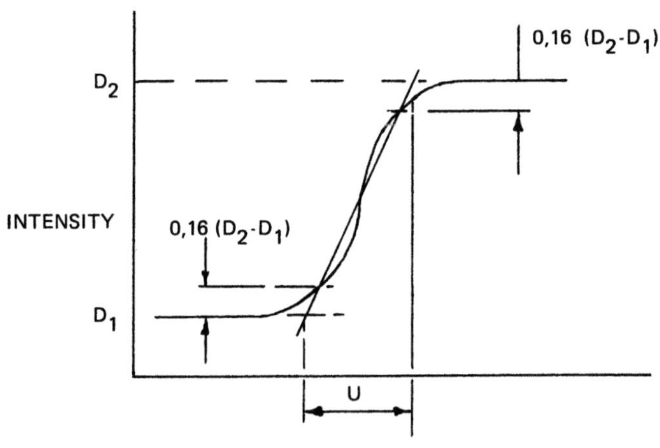

B. KLASENS' METHOD

Fig. 1.13 Measurement of Unsharpness.

There are three types of unsharpness in a radiography system: geometric, foil and film. Geometric unsharpness was discussed earlier and is readily determined from the collimator dimensions and the size of the sample but the other two are difficult to seperate and must be assessed together by the Klasens technique. The combined effect of different types of unsharpness was also considered by Klasens, who evolved the following form of empirical relationship

$$U_t = (U_g^{\,3} + U_{ff}^{\,3})^{1/3} \qquad (15)$$

where U_t = total unsharpness
U_g = geometric unsharpness
U_{ff} = foil and film unsharpness

Thus the source of unsharpness cannot be considered in isolation and their interdependence shows that there is little point in achieving a small geometric unsharpness if the foil/film unsharpness is much larger.

Whilst the above methods of measuring unsharpness are widely used in all forms of radiography they fail to acknowledge that resolution is a function of image contrast for they only give the resolution at one particular contrast value. The important failing of this system is that is does not allow the radiographer to systematically evaluate his radiography system to determine the conditions under which it will produce the maximum information. A system which comes much closer to this ideal has been in use by designers of communication and optical systems for many years and has been advocated for use in radiography for about a decade. This is a frequence response method of evaluating the performance of a system, usually called the Modulation Transfer Function.

It can be shown that any image, even a series of step changes in intensity, may be expressed as a series of sine waves of differing frequency, amplitude and phase by the use of Fourier analysis. It follows therefore that a test-object which will transmit an intensity pattern which varies sinusoidally with known frequency and amplitude would provide a means of evaluating the detail-recording capability of a radiographic system. Such a test-object is shown at Figure 1.14 and ideally the radiographic system should reproduce an image of such an object without loss or change of information.

In practice such test objects are difficult to produce and so the required information is obtained by the use of a narrow slit, or more conveniently, a sharp edge. This, of course, brings us back to the type of curve shown in Figure 1.13, and by analysing this curve into its component sine waves the type of curves shown in Figure 1.15 can be produced to show how the resolving power varies with contrast. The resolving power is expressed as the object spatial frequency and this can be conveniently regarded as the maximum number of lines per millimeter, that can be resolved. For example, a spatial frequency of 2 cycles per mm would be equivalent to a resolution of two lines per millimeter (two ¼ mm wide lines seperated by a ¼ mm gap). The figure shows that the gadolinium foil will give the best resolution over the entire contrast range that was measured,

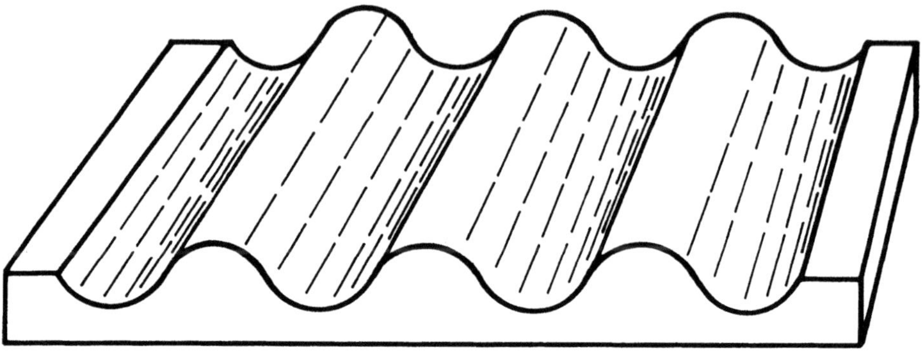

Fig. 1.14 Diagram of a Test Object whose Transmission varies Sinusoidally along its Length.

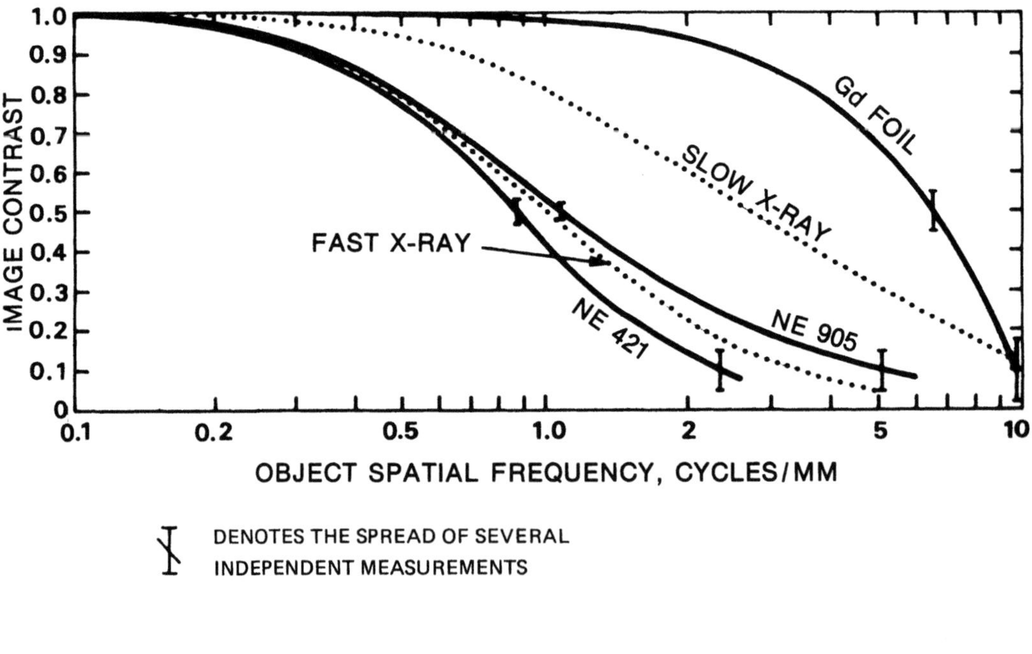

Fig. 1.15 Contrast Transfer Functions of NE241 and NE905 Scintillators, Gadolinium Foil and X-Ray Films.

only being equalled by the slow X-ray film at the contrast of about 0.1. This resolution is at the expense of speed, in general the higher the resolving power the slower the detector.
Whilst this method has considerable advantages over the methods described earlier, it is time-consuming to apply and so it is mainly used by experimentalists and designers who wish to analyse a complete radiographic system. Modulation transfer functions of individual components in a system i.e. the collimation, converter foil, photographic film etc. can be measured separately and then multiplied to give the overall modulation transfer function. Such a technique is a great aid to design, in that the weakest part of the system can be evaluated and improvements made where they will be most effective. For further information on the frequence responce method see Halmshaw [Ref. 42] and Halms [Ref. 43].

1.1.8.3 Some Observations on Resolution Sensititvity

Radiography is commonly used to detect voids and inclusions within a material and so it is often necessary to know the minimum size of void that can be detected.
Let us ask the question 'Can the detector/recorder combination discriminate between the beam intensity which will form the image of the bulk of the sample and the intensity which will form the image of that part of the sample which contains the void.' The answer will depend upon the response of the detector/recorder being used. With a metal foil in combination with a photographic film the photographic response is linear; when the combination is a light-emitting scintillator and a film then the response will be logarithmic. Consider these two cases.
The relationship between the incident neutron intensity and the transmitted neutron intensity (ignoring scattering) will be

$$\phi = \phi_o e^{-\Sigma x} \qquad (16)$$

where ϕ = transmitted intensity, n $cm^{-2}s^{-1}$
ϕ_o = incident intensity, n $cm^{-2}s^{-1}$
Σ = total macroscopic cross section, cm^{-1}
x = thickness of sample, cm

Assuming we are using a metal foil then the film density (De) will have a linear relationship with the neutron exposure, i.e.

$$D_e \sim \phi \qquad (17)$$

and we can rewrite (16) as

$$D_i = D_b e^{-\Sigma x} \qquad (18)$$

where D_i = image density of the film
D_b = background density of the film

Differentiating and expressing as small differences gives:

$$\Delta D_i = D_b \Sigma \cdot \Delta x \qquad (19)$$

where ΔD_i = minimum detectable density change
Δx = minimum dectectable thickness change in the object, cm

For a metal foil detector

$$D_e = G \cdot E \qquad (20)$$

where G = slope of density / exposure curve for the film
E = exposure, n cm^{-2}

So, from (19) and (20)

$$\Delta x = \frac{\Delta D_i}{G \cdot E \cdot \Sigma} \qquad (21)$$

This equation shows that in order to detect the smallest thickness change, or detect the smallest void, then the film must have the highest contrast (G) and the exposure must be as high a possible.
Now the second case is where the metal screen is replaced by a light emitting scintillator, and for this there will be logarithmic relationship between film density and neutron exposure, i.e.

$$D_e = G\,(\log E) \qquad (22)$$

or

$$D_e = G\,(0{,}434 \ln E) \qquad (23)$$

differentiating and expressing as small differences gives

$$D_e = 0{,}434 \; \frac{G \cdot \Delta E}{E} \qquad (24)$$

Now as in equation (19) the fractional difference in exposure can be expressed as

$$\frac{\Delta E}{E} = \Sigma x \qquad (25)$$

and so (25) becomes

$$\Delta D_e = 0{,}434 \cdot G \cdot \Sigma \cdot \Delta x \qquad (26)$$

or

$$\Delta x = 2{,}3 \frac{\Delta D_e}{G \cdot \Sigma} \qquad (27)$$

where

ΔD_e = minimum detectable density change
Δx, G, Σ = as above

Hawkesworth [Ref. 23] considered the above sensitivity equations and showed that for ionising radiation (i.e. metal foils) the background radiation incident on the film during the exposure has no effect on the thickness sensitivity. For light emitting screens however the thickness sensitivity is dependent upon the background exposure and equation (27) becomes

$$\Delta x = \frac{2{,}3 \cdot D}{G \cdot \Sigma} \; \frac{1 + E_o + D_b}{E} \qquad (28)$$

where D_b = background exposure

E_o = intercept of the film characteristic curve on the abscissa – see Figure 1.14.

This leads to the desire to maximise the exposure to achieve the highest thickness sensitivity, but is must be said that this may be nullified by the consequent increase in the background exposure caused by the gamma radiation in the beam.

We now have two expressions which give a reasonable description of the minimum detectable void, or the minimum detectable thickness of a non-scattering material in a neutron beam which is free from interfering radiation such as fast neutrons and gamma rays. The allowance which must be made when scattered neutrons and gamma rays are present depends upon the type of detector in use. If the transfer technique is being used then the gamma rays arriving at the detector will not be recorded, and similarly some of the scintillators (direct technique) have a sufficiently low gamma sensitivity for this factor to be ignored. If a gadolinium foil (direct technique) is being used then all of these radiations will be present and will contribute to a reduction in the contrast-sensitivity.

Furthermore thermal neutrons are attenuated by matter in two ways, they are absorbed or scattered. In the first case they are removed from the beam and in the second they are scattered in all directions from the target nuclei. So to the detector the first process in unequivical, the neutron is removed from the beam and is simple not there to be recorded. The response to scattering however depends upon the spacing between the scattering material and the object. If they are widely spaced the most of the attenuated neutrons will be scattered out of the beam and once more will not be there to be recorded. This 'widely spaced' geometry is, in principle, similar to that used when cross section measurements are made and so the equations which use cross section values will only be accurate for scattering materials when the object and the detector are spaced apart. However, the usual practice in neutron radiography is to place the object and the detector as close as possible, and so many of the scattered neutrons are recorded.

Gamma rays are also scattered and absorbed by the sample but their absorption is usually negligible at the gamma ray energies which predominates in the beam. The effect of gamma ray scattering on the radiograph is similar to that of scattered neutrons in that they reduce the contrast. The gamma ray component of the beam is always made as small as possible and so gamma ray scattering is not usually a serious problem (unless a gamma emitting specimen is being radiographed), although a direct technique neutron radiograph can usually be improved by placing a lead or bismuth filter in the beam.

A more detailed examination on the detection of voids is given in Appendix 1.4.

1.1.9 Neutron Beam Filters

Neutron radiography is usually performed with a neutron beam containing a range of energies, but the predominant value will be the thermal energy ($\sim$ 0,025 eV). Now although the converter foils are generally most sensitive at the thermal energy, and the response falls steadily as the energy increases, superimposed upon this effect are the large resonances which occur at well-defind energies. At these energies the cross-section of the foil increases/decreases rapidly, often by several orders of magnitude, and then generally returns to about its previous value.
So if a filter is placed in the neutron beam which will only pass neutrons whose energy coincides with the resonance in the converter foil then the system will be more sensitive to the remaining neutrons. This ideal situation can be approached by using for example, a cadmium filter and an indium converter foil. The cadmium will absorb all neutrons of energy below about 0.4 eV and so leave the beam relatively rich in higher energies. One of these, 1.4 eV, is the energy of a resonance in indium at which the cross-section increases from 2×10^2 barns to 3×10^4, and an example of the use of this particular filtered neutron energy is found in the radiography of fuel pins where the cross-section of uranium is lower at 1.4 eV than at thermal energy and so a fuel pin is more readily penetrated by neutrons at this higher energy. Spowart [Ref. 33] investigated this effect and found that the penetration of a 0.6 cm diameter (20% Pu, U^{235}-enriched, mixed-oxide pellet) was improved by about 40 x by using 1.4 eV neutrons. It must be recognised that this technique works because the thermal neutrons, which are normally responsible for most of the film blackening, have been removed from the beam. With a non-filtered beam these would fully expose the image before the more penetrating neutron could make a perceptible contribution. The filter therefore usually reduces the beam intensity and a longer exposure may be required. This effect will vary with the neutron energy, and the materials of the foil and the sample, but taking the fuel pin example given above we can say that the increase in exposure time will be a function of:

(a) the ratio of 1.4 eV neutrons to thermal neutrons in the unperturbed neutron beam,
(b) the decrease in neutron intensity produced by the filter,
(c) the change in transmission through the sample, and
(d) the increased sensitivity of the foil.

The effect of these factors will, of course, vary with each beam and sample combination. Further examples of this technique are discussed by Miller and Watanabe [Ref. 38] who used a combined cadmium/gadolinium/indium filter to detect the 4.3 eV resonance in tantalum. Fig. 1.16 shows the transmission of the filter, the resonance in the gold detector and the object. They also discussed the use of a high purity silicon filter to penetrate hydrogen.
Cold neutrons [Ref. 19] have also been used to increase the neutron penetration of crystalline materials and Fig. 1.8 shows the fall in neutron attenuation coefficient for aluminium, iron, zirconium, tin and lead, compared to say hydrogen and gadolinium, where the attenuation coefficient increases.
A list of potential filter materials is given in Table 1.6.

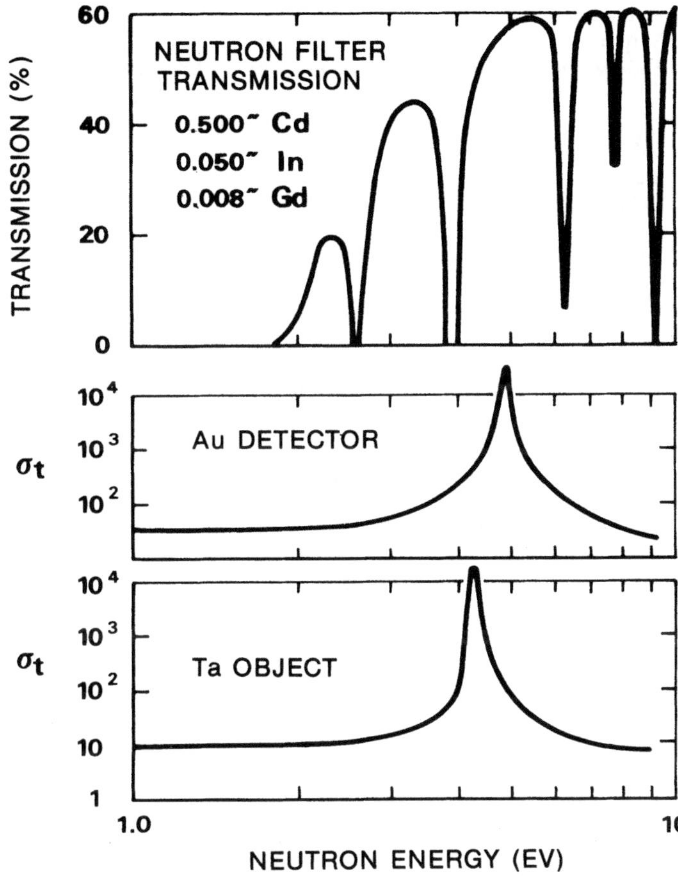

Fig. 1.16 Resonance Curves for a Cd/In/Gd Filter, a Gold Foil and a Tantalum Sample.

Table 1.6 Neutron Filters

Element	Minimum Energy, eV	Absorption Cross-section Barns	Maximum Energy, eV	Absorption Cross-section Barns	Temp. °K	Resonance Type 1)
Beryllium 4	$5x10^{-3}$	0.45	$7x10^{-3}$	6	300	A
Beryllium 4	$3.5x10^{-3}$	0.05	$7x10^{-3}$	6	100	A
Beryllium Oxide	$2x10^{-3}$	2	$6x10^{-3}$	10		A
Boron 5	$6.5x10^{-4}$	$4.5x10^{3}$	$3x10^{3}$	6		C
Carbon 6	$1.5x10^{-3}$	0.55	$2x10^{-3}$	8		A
Sodium 11	$5x10^{2}$	3.2	$3x10^{3}$	380		D
Silicon 14	$1.4x10^{5}$	0.45	$1.9x10^{5}$	12		E
Sulphur 16	$7x10^{4}$	0.45	$1.2x10^{5}$	20		E
Scandium 21	30	20	$4x10^{3}$	40		G
Iron 26	$2.6x10^{4}$	0.45	$2.9x10^{4}$	3(av.)		E
Rhodium 45	1.3	$5x10^{3}$	20	4.5		D
Rhodium 45	0.5	100	1.0	10		D
Cadmium 48	0.18	$8x10^{3}$	10	4.5		B
Xenon 54	0.1	$3x10^{6}$	1.0	10^{4}		B
Promethium	3.5	40	5.5	$2.5x10^{4}$		F
Gadolinium 64	$2x10^{-3}$	$1.4x10^{5}$	1.5	35		B
Bismuth 83	$6x10^{-4}$	0.55	$2x10^{-3}$	8	300	A
Bismuth 83	$8x10^{-4}$	0.3	$8x10^{-3}$	1	100	A
Proactinium 91	$0.4x10^{5}$	0.05	$1.5x10^{6}$	1		A
Plutonium 94	0.3	10^{3}	6.0	20		F

1) KEY TO RESONANCE TYPES

1.1.10 Tomography

Tomography is a radiographic technique first developed for medical X-radiography in which a series of exposures are taken at uniform intervals around a specimen and, by the use of a computer programme, the data is transposed to give a picture of a cross section a right angles to the plane of the radiographs.

Barton [Ref. 17] et al has used epicadmium neutron radiography with this method in order to examine nuclear reactor fuel bundles having 217 pins in a hexagon array, requiring the penetration of 9 fuel pins having a total attenuation of 47 cm^{-1}.

The computer output is in the form of a 'dot-picture' in which the sensivity is limited by the dot size, but at the moment the system resolution is in the order of a few millimeters and so this is not a limitation. However, a resolution of a few millimeters is, by normal standards, very poor, but even this is much better than no information at all from samples that are normally extremely difficult to penetrate.

1.2 THE DESIGN OF NEUTRON RADIOGRAPHY EQUIPMENT

Having considered the principles relating to neutron radiography we can now examine some of the topics discussed earlier in greater depth in order to design or select a neutron radiography unit suitable for the radiography of a particular range of samples.

1.2.1 The Choice of Neutron Source

The starting point is to ask a few questions, viz
- is the object radioactive,
- what resolution do you require,
- can the object be taken to the equipment or must the equipment go to the object,

in order to determine wheter a reactor or an accelerator/isotopic source is to be used. Reactor sources are undoubtedly the most powerful of those discussed earlier in this text and if high resolution is required then a reactor source must be used. Its disadvantages are that it is immobile and expensive. A small reactor would cost $ 500,000 or more, it requires a licence to operate and a small staff to out those operations. On the plus side of the argument can be added the fact that it would provide a larger number of neutron beams than the accelerator and isotopic assemblies, the flux is more stable and the cost per neutron is much less. Typically a nuclear reactor will provide neutron fluxes at the collimator inlet that are some 10^3 to 10^6 times higher than those from the alternative sources.

Let us assume that your resolution requirements are sufficiently exacting to require the use of metal foil converter and that you require to locate the neutron radiography equipment in your works to examine a variety of components. Your company has radiography department who's staff are familiar with high energy X-ray equipment and so you advise the management to obtain a 3 MeV Van der Graaff accelerator with a beryllium target of the type shown in Figure 17.
This arrangement will produce neutrons with an energy of about 5 MeV and this can be moderated by inserting the accelerator tube into the centre of a block of polythene or a tank of water.

If you now ask for the assistance of the Physics Department in your organisation to determine the magnitude and position of the peak thermal flux they will take the details of the flux spectrum from the accelerator target and perform the necessary diffusion theory calculations. The peak thermal flux will probably be about 20 cm from the target and the thermalisation factor i.e.

$$\frac{\text{fast neutron flux at the target, n cm}^{-2}\text{s}^{-1}}{\text{peak thermal neutron flux in the moderator, n cm}^{-2}\text{s}^{-1}}$$

will be about 100. If we assume that your accelerator produces a source intensity of about 4×10^{11} n s^{-1}, this will give a peak thermal flux in the moderator of 4×10^{9} n $cm^{-2}s^{-1}$.

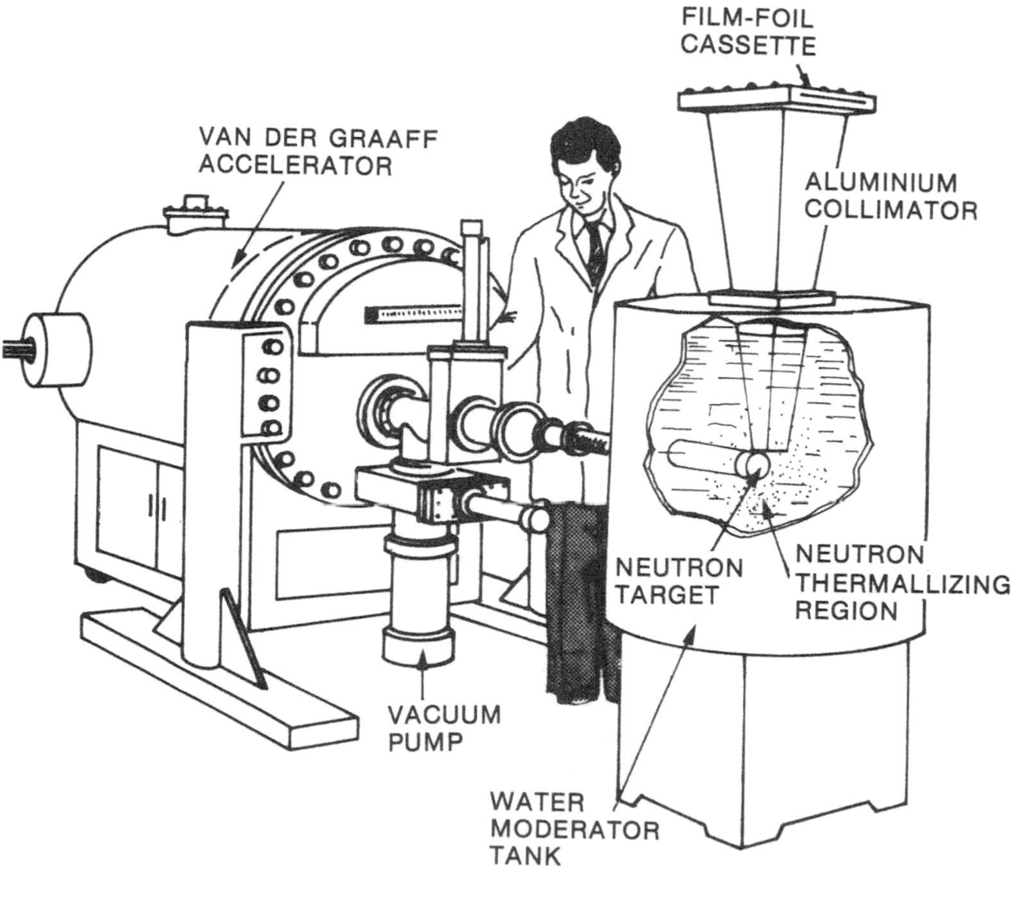

Fig. 1.17 Van Der Graaff Accelerator for Neutron Radiography.

1.2.2 The Collimator

1.2.2.1 Collimator Design

The next stage will be to consider the collimation, and the neutron flux required at the foil. Your objects will not be radioactive and so the direct technique can be used, and, as the Inspection Department require a resolution of (say) 0.01 cm you are going to use a gadolinium converter foil. As the incident neutrons will all be absorbed in about a 10 μm thickness of gadolinium then the foil can be thin and a 25 μm thickness mounted on an aluminium backing plate would be a practical choice.

As the photographic film will be placed in the neutron beam with the foil it will respond to the gamma-rays in the neutron beam. The gamma exposure will reduce the contrast so you will probably need to place a lead or bismuth filter, say 0.5 cm thick, across the entrance to the collimator to absorb the unwanted gamma's.

The L/D ratio of the collimator can be determined from the equation (29) i.e.

$$U_g = \frac{D}{L_s} L_f \qquad (29)$$

where U_g = geometric image unsharpness = 0,01 cm
D = source-aperture size
L_f = image-to-object distance = 0,5 cm
L_s = source-to-object distance

if we make the assumption that
$L_s = L =$ collimator length

the equation now becomes

$$\frac{L}{D} = \frac{L_f}{U_g} \qquad (30)$$

and so knowing that L_f is normally equal to the thickness of the object and given that U_g is known, then L/D can be determined.

The next problem to be considered is the selection of a material to line the walls of the collimators choosing from the list given earlier, i.e. boron (in the form of boral), cadmium, dysprosium, europium, gadolinium or indium.

The nuclear effectiveness of these materials can be assessed from Fig. 1.6. This shows the total cross-section plotted against the neutron energy, and the absorption-to-scattering cross-section ratio for thermal neutrons.
The effectiveness of an absorbing material will vary with the neutron-energy spectrum of the neutron beam (compare the cross-section of Cd and In at 10^{-2} and 1.4 eV), and this spectrum can be broadly characterised by the cadmium ratio* of the beam. Fig. 1.6 shows that on the left-hand side of the cadmium cut-off line the most effective materials are europium, gadolinium and cadmium. All of these materials have a high absorption-to-scattering cross-section ratio, which

means that there is a high probability that a neutron will be absorbed in the lining rather than being scattered into the beam.

Cadmium has a disadvantage in that it emits a high energy gamma ray when a neutron reaction occurs and this will add to the unwanted radiation in the beam.

The information on the-right hand side of the cadmium cut-off line indicates that, for the higher-energy neutrons, indium is generally the most effective, closely followed by all of the other materials except cadmium.

It is thus clear that there are no outstanding materials from the neutronic viewpoint and so the cost and the mechanical properties of these materials must also be considered when making a selection.

The cost of each of these materials is clearly something which could vary considerably with time so these must be determined at the time of need. The properties of interest for boron, cadmium and europium are given in the Tables 1.7, 1.8 and 1.9 of the following section and those for indium, dysprosium and gadolinium will be found in the Section of Characteristics of Foil Materials (see 1.2.3.1), Tables 1.10, 1.11 and 1.12.

1.2.2.2 Characteristics of Lining materials

Boron is a light metal of high hardness and melting point which is normally made into solid shapes by powder metallurgy techniques. It can be obtained in the forms of boron carbide, boron oxide, boron nitride and boral. This latter form is a mixture of boron carbide and aluminium which is clad in aluminium.

Boral is somewhat difficult to machine but can be sawn, sheared or spark eroded. A 6 mm thick sheet can be rolled to a minimum diameter of about 200 mm.
When used as a converter foil the material is in the form of enriched ^{10}B vacuum deposited upon aluminium, or as a boron powder in a plastic matrix.

Table 1.7 shows the natural material consists of 20% B^{10} and 80% B^{11} and that under neutron irradiation the B^{10} is converted to $L_1{}^7$ with a production of a 2.3 MeV alpha particle which is responsible for the damage tracks in nitrocellulose when the track etch method is used.

*) The ratio of two neutron flux measurements made by irradiating a foil or a wire with and without a cadmium cover. Such a cover is taken as giving a cut-off at 0.4 eV.

Table 1.7 Data on Boron [Refs. 24,25,26]

General

Appearance	Depends upon the form	
Density	2.45 g cm^{-3}	
Atomic Number	5	
Atomic Weight	10.82	
Atomic Density	1.36×10^{23} atoms cm^{-3}	
Natural Cross Sections:	Microscopic	Macroscopic
Absorption	755×10^{-24}cm	$103cm^{-1}$
Scattering	4×10^{-24}cm	0.346 cm^{-1}
Total	759×10^{-24}cm	$103cm^{-1}$

Isotopic Composition

Isotopic Number	Abundance %	Half Life	Emission		Transmutation	
			Type	Energy MeV	Cross Section Barns	Isotope formed
10	18.3	Stable	alpha	2.3	3836	Li^7
11	81.7	Stable	n, γ	13.4	0.005	B^{12}

Cadmium is readily available in foil and sheet form from about 0.04cm thick. It is a very soft metal with a dull mottled appearance and the surface usually has imperfections. When used as a converter foil the surface should be polished until it is smooth and flat.

It is readily fabricated by all of the commercial techniques. It oxidises very slightly in air and does not react with boiling water.

Table 1.8 shows that 98% of the natural material consists of six isotopes, Cd^{110}, Cd^{111}, Cd^{112}, Cd^{113}, Cd^{114} and Cd^{116} of which only the $Cd^{113}(n,\gamma)Cd^{114}$ has a large cross section. This reaction produced a gamma ray of 9 MeV which is the radiation mainly responsible for the blackening of film when cadmium is used as a converter foil. When cadmium is used as a lining then this gamma ray will enter. The neutron beam and cause a reduction in contrast or fogging.

Table 1.8 Data on Cadmium [Refs. 24, 25, 25]

General

Appearance	Soft metal, dull surface	
Density	8.65 g cm^{-3}	
Atomic Number	48	
Atomic Weight	112.41	
Atomic Density	4.64×10^{22} atomic cm^{-3}	
Natural Cross Sections:	Microscopic	Macroscopic
Absorption	$2{,}450 \times 10^{-24} cm^2$	114 cm^{-1}
Scattering	$7 \times 10^{-24} cm^2$	0.325 cm^{-1}
Total	$2{,}457 \times 10^{-24} cm^2$	114 cm^{-1}

Isotopic Composition

Isotopic Number	Abundance %	Half Life	Emission Type	Emission Energy MeV	Transmission Cross Section Barns	Transmission Isotope Isotope Formed
106	1.2	Stable	n, γ		1.0	Cd^{107}
107		6.5h				
108	0.9	Stable	n, γ		1.1	Cd^{109}
109		453d	n, γ		650	Cd^{110}
110	12.4	Stable	n, γ		11	Cd^{111}
					0.1	Cd^{111m}
111	12.8	Stable	n, γ		24.3	Cd^{112}
111m		49m	gamma	0.245		
112	24.0	Stable	n, γ		0.06	Cd^{113}
113	12.3	Stable	n, γ			Cd^{114}
113		14.6y	Beta	0.6		In^{114}
113		9×10^{15}y	Beta	0.3	19,910	In^{114}
114	28.8	Stable	n, γ		0.036	Cd^{115}
115		44.8d	Beta	1.6		
115m		53.38h	Beta	1.1	0.027	Cd^{117}
116	7.6	Stable	n, γ		0.050	Cd^{117m}

Europium is a rare earth material that is of interest as a reactor control material because, under neutron irradiation, it produces a series of high cross section daughter-products which cause the initial cross section of 4300 barns for the natural material to fall to about 700 barns and then stay relatively constant even at high irradiation densities.

It is available as europium oxide and as a dispersion (probably about 12%) in titanium. It oxidises rapidly in air at room temperature and should therefore be handled and stored under an inert atmosphere. Finely divided europium can ignite spontaneously in air. It also reacts vigorously with cold water.

Table 1.9 shows that the natural material consists of 47.8% Eu^{151} and 52.12% Eu^{153} with two large cross sections for the Eu^{151} (n, γ) Eu^{152} reactions.

Table 1.9 Data on Europium [Refs. 24, 25, 26]

General

Appearance		
Density	5.22 g. cm^{-3}	
Atomic Number	63	
Atomic Weight	152	
Atomic Density	2.07 x 10^{22} atoms cm^{-3}	
Natural Cross Sections:	Microscopic	Macroscopic
Absorption	4300 x $10^{-24}cm^2$	89 cm^{-1}
Scattering	8 x $10^{-24}cm^2$	0.166 cm^{-1}
Total	4308 x $10^{-24}cm^2$	89.2 cm^{-1}

Isotopic Composition

Isotopic Number	Abundance %	Half Life	Emission Type	Emission Energy MeV	Transmutation Cross Section Barns	Transmutation Isotope Formed
151	47.8	Stable	n, γ		3300, 5900, 4.0 1)	Gd^{152}
152		9.3h	beta	1.9		Gd^{152}
152		12.4y				
152m		96m	gamma	0.04,0.09		Eu^{153}
153	52.12		n, γ		390	Eu^{154}
154		8.5y	Beta	0.6,1.8		Gd^{154}
155		4.96y	Beta	0.1,0.2		Gd^{155}
156		5.2d	Beta	0.5,2.4		Gd^{156}
157		15.15h	Beta	1.3		Gd^{157}
158		46m	Beta	2.4,3.4		Gd^{158}
159		18.7m	Beta Gamma	2.6 0.045,1.76		Gd^{159}
160		42s	Beta	3.9		Gd^{160}

1) The three cross section values belong to different states of the ^{151}Eu nucleus.

1.2.2.3 Defining the Inlet Aperture

As the resolution of the collimator is a function of the inlet-aperture size it is important that this aperture should be well defined. This can be achieved by constructing the inlet face of the collimator from a material which is opaque to neutrons, and especially those neutrons to which the converter foils are most sensitive. This leads to the conclusion that the inlet face should be made from layers of the converter foil materials that will be used.

The most common of these are dysprosium, gadolinium and indium, but further examination of Fig. 6 shows that gadolinium has the highest cross section values on the left of the cadmium cut-off line and that indium predominates on the right hand side, so that a combination of these two materials only would probably provide an acceptable front face for the collimator.

The appropriate thicknesses can be calculated from:

$$A = 1 - e^{-\Sigma d} \qquad (31)$$

where

A = attenuation factor
Σ = macroscopic cross-section, cm^{-1}
d = thickness, cm

The thickness should be chosen to make the attenuation factor equal to at least 0.95 and the cross-section values appropriate to about 5 eV should be used in order to ensure that all of the neutrons of energies to which the foils are sensitive are absorbed in the inlet face.

1.2.2.4 Divergence Angle

In travelling from the source aperture to the foil the neutrons in a divergent collimator will follow a shorter path at the collimator centreline than at the walls, assuming the target is a plane surface. As the neutron flux will vary with the square of the collimator length then clearly the dose at the centre of the foil will be greater than that at the edges. However, this only a problem for low flux neutron radiography units with short and wide collimators, for it can be shown that it requires a divergence angle of 35° to produce a 10% exposure difference between the centre and the edge of a foil.

1.2.2.5 Geometric Enlargement and Diminution

The divergence of the beam will cause the radiographic image to be generally larger or smaller than the object size, depending on the relative sizes of the object and the size of the inlet aperture to the collimator. The geometry is shown in Fig. 1.7D and the percentage change in the height of the object as recorded at the image is

$$\frac{L_f}{L_s} \left(1 - \frac{D}{h_o}\right) 100\% \qquad (32)$$

1.2.3 The Converter Foils

1.2.3.1 Characteristics of Foil Materials

There are a considerable number of materials which have been tested as converter-screen Materials [Ref. 4], but the foils that are now in general use are indium and dysprosium for the transfer method using X-ray films and gadolinium for the direct method also using X-ray films and boron and lithium foils for the track etch method. The characteristics of these materials are as follows.

– Characteristics of Indium

Indium metal is readily available as a foil from about 0.05 to 0.1 cm thick. It is a soft metal and has a dull mottled appearance when received, and the surface usually has slight undulations. This can cause small density variations on the radiographs and so it is advisable to polish the surface until it is smooth and flat.

Table 1.10 shows that the natural material consists of 4.3% In^{113} and 95.7% In^{115}. The cross-section of the In^{113} isotope for conversion into In^{114} is small, so very little In^{114} is produced (and this only has a half life of 72 seconds) and only one of its two alternative states of In^{115} has a significant cross-section for conversion into In^{116}.

The In^{116} isotope is the most important for the transfer method because it emits a 1.0 MeV (maximum) beta ray when it decays to Sn^{116}. This decay has a half-life of 54 minutes, and so allows reasonable neutron exposure and transfer times.

An alternative mode of decay for In^{116} has a 14-second half-life and emits a 3.3 MeV beta. This will contribute to the exposure of the photographic film, but in practice the short half-life makes it difficult to use this radiation since it is emitted during the time while the foil is being transported from the exposure position to the darkroom. Anyway, owing to its high energy it will be a low-resolution contribution [Ref. 27].

The isotopes of indium beyond In^{116} do not have appreciable cross-sections for neutron absorption and so they do not make any practical contribution to the process of neutron-radiographic image formation.

Table 1.10 Data on Indium [Refs. 24, 25, 26]

General		
Appearance	Soft metal, dull surface	
Density	7.28 g cm^{-3}	
Atomic Number	49	
Atomic Weight	114.82	
Atomic Density	3.82×10^{22} atoms cm^{-3}	
Natural Cross-Sections:	Microscopic	Macroscopic
Absorption 1)	196×10^{-24} cm^2	7.75 cm^{-1}
Scattering	2.2×10^{-24} cm^2	0.084 cm^{-1}
Total	198.2×10^{-24} cm^2	7.564 cm^{-1}

Isotopic Composition

Isotopic Number	Abundance, %	Half-Life	Emission Type	Emission Energy, MeV	Transmutation Cross-Section, Barns	Transmutation Isotope formed 2)
113	4.3	Stable	n, γ		3.9	In^{114}
					7.5	In^{114m}
114		71.9 s	beta	2.0		Sn^{114}
114m		49.5 d	gamma	0.192		
115	95.7	Stable	n, γ		155	In^{116}
115		6×10^{14}y	beta	0.5	65	In^{115}
					45	In^{115}
					92	In^{115m}
115m		4.5h	beta	0.8		
116		54m	beta, gamma	3)		In^{117}
116		14s	beta	3.3		Sn^{116}
116m		2.2s	gamma	1.64		
117		38min	beta	0.7		Sn^{117}
117m		1.95n	beta	1.8		

1) This is a reaction, which determines the activity of the foil after irradiation and hence the neutron image.
2) m = metastable state
3) β: 0.34 (1.5 %), 0.59 (11 %), 0.87 (40 %), 1.0 (49 %)
γ: 0.138 (3 %), 0.147 (36 %), 0.819 (17 %), 1.09 (53 %), 1.293 (80 %), 1.508 (11 %), 2.111 (20 %).

Dysprosium metal is obtainable as foil up to about 0.025 cm thick. It has a semi-bright, smooth appearance and is sufficiently hard to withstand normal handling without incurring scratches or abrasions which will show on a radiographic image.

Table 1.11 shows that naturally occuring dysprosium has seven stable isotopes, of which Dy^{156} and Dy^{158} can be neglected owing to their small abundance. The isotopes Dy^{160}, Dy^{161}, Dy^{162} and Dy^{163} are not important for the transfer method since they do not form radioactive isotopes. The important isotope is Dy^{164} since this has a large cross-section for the formation of Dy^{165}, which is formed with a half life of 2.35 hrs and which decays into stable Ho^{165} with the emission of 1.3 MeV (maximum) beta rays. This transition has a 2.35 hr half-life. Dy^{165} also has a metastable state, and this has an associated decay emission of 1.0 MeV (maximum) beta rays, but the half-life is only 1.3 minutes and in practice this isotope does not contribute much to the photographic exposure.

The isotopes of dysprosium beyond Dy^{165} do not have any great cross-section for neutron absorption and so do not make any practical contribution to the process of neutron-radiographic image formation.

Table 1.11 Data on Dysprosium [Refs. 24, 25, 26, 50]

General

Appearance	Hard metal, semi-bright surface	
Density	8.56 g cm^{-3}	
Atomic Number	66	
Atomic Weight	162.51	
Atomic Density	3.17×10^{22} atoms cm^{-3}	
Natural Cross-Section:	Microscopic	Macroscopic
Absorption	950×10^{-24} cm^2	28.9 cm^{-1}
Scattering	100×10^{-24} cm^2	3.17 cm^{-1}
Total	1050×10^{-24} cm^2	33.4 cm^{-1}

Isotopic Composition

Isotopic Number [1]	Abundance %	Half-life	Decay Emission Type	Decay Emission Energy, eV	Transmutation Cross-Section, barns	Isotope formed [1]
156	0.05	Stable	n, γ		33	Dy^{157}
157		8.1 h				Tb^{157}
158	0.09	Stable	n, γ		96	Dy^{159}
159		144.4d				Tb^{159}
160	2.29	Stable	n, γ		55	Dy^{161}
161	18.88	Stable	n, γ		600	Dy^{162}
162	25.3	Stable	n, γ		160	Dy^{163}
163	29.7	Stable	n, γ		125	Dy^{164}
164	28.18	Stable	n, γ		2590	Dy^{165}
					1700	Dy^{165m}
165		2.35 h	beta, gamma	[2]	3900	Ho^{165}
165m		1.3 m	beta	0.9		
166		81.5 h	beta	0.4,0.5		Ho^{166}

1) m = metastable state.
2) beta : 0.22 (0.1 %), 0.254 (0.03 %), 0.3 (1.3 %), 1.2 (15 %), 1.3 (83 %).
gamma : 0.094 (10 %), 0.279 (1 %), 0.361 (40 %), 0.71 (2 %), 1.02 (8 %).

Gadolinium metal is obtained as a foil up to about 0.025 cm thick. It has a bright smooth appearance and it is strong enough, as a foil, to withstand normal handling without incurring scratches or abrasions which will show on a radiographic image.

Table 1.12 shows that naturally occuring gadolinium has six stable, and one very long lived, isotopes of which Gd^{152} and Gd^{154} can be neglected owing to their small abundance. The important isotopes are Gd^{155} and Gd^{157} since these have large cross sections.

Table 1.12 Data on Gadolinium [Refs. 24, 25, 26]

General

Appearance	Hard metal, bright surface	
Density	7.95 g. cm^{-3}	
Atomic Number	64	
Atomic Weight	157.26	
Atomic Density	3.05 x 10^{22} atoms cm^{-3}	
Natural Cross-Sections:	Microscopic	Macroscopic
Absorption	46,000 x $10^{-24}cm^2$	1,403 cm^{-2}
Scattering	–	–
Total	–	–

Isotopic Composition

Isotopic Number	Abundance %	Half Life	Emission Type	Emission Energy MeV	Transmutation Cross Section Barns	Transmutation Isotope Formed
152	0.2	$1.1x10^{14}$y	alpha	2.14	1100	Eu^{149}
153		241.6d	gamma	0.097, 0.103		
154	2.2	Stable	n, γ		85	Gd^{155}
155	14.9	Stable	n, γ		61,000	Gd^{156}
156	20.6	Stable	n, γ		1.5	Gd^{157}
157	15.7	Stable	n, γ		254.000	Gd^{158}
158	24.7	Stable	n, γ		2.5	Gd^{159}
159		18.56h	Beta	0.9		Tb^{159}
160	21.7	Stable	n, γ		0.77	Gd^{161}
161		3.6m	Beta	1.6		Tb^{161}
162		8.2m	Beta	1.0		Tb^{162}

1.2.3.2 Converter-Foil Thickness and Speed.

The choice of the converter-foil thickness is a compromise between

(a) a thin foil for high resolution and

(b) a thick foil for short exposure times and sufficient rigidity for handling.

The choice is limited by the thickness of foils available and by the difficulties of handling extra-thin foils.

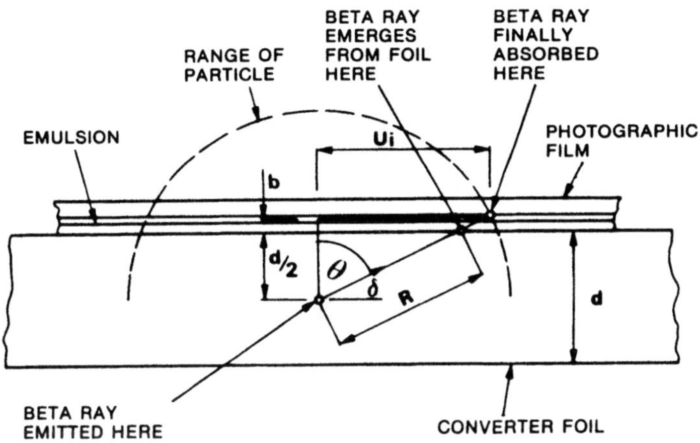

Fig. 1.18 Film Foil Geometry.

Fig. 1.18 is a sketch of the film and foil combination and shows the path of a particle which has been emitted from an atom in the centre of the foil. For any particular direction of emission towards the film, a simple theoretical relationship would be

$$\tan \mu = \frac{U_i}{b + d/2} \tag{33}$$

$$\text{i.e. } U_i = (b + d/2)\cdot\tan \mu,$$

where

U_i = inherent unsharpness, cm
θ = angle of emission, degrees
b = distance from film surface to emulsion, cm
d = foil thickness, cm.

So theoretically the unsharpness is directly proportional to the film and foil dimensions, and the smaller these are the smaller the unsharpness becomes. The activity on the foil per unit thickness is given by

$$\frac{S}{d} = \Sigma\phi\,(1 - e^{-\lambda T}), \tag{34}$$

where

S = activity, disintegrations s^{-1} cm^{-2}
Σ = cross-section, cm^{-1}
ϕ = neutron flux, n cm^{-2} s^{-1}
d = foil thickness, cm
T = irradiation time, s
λ = decay constant 0,69/
τ = half-life of foil material, s.

Now, ignoring self-shielding, equation (34) shows that for a particular irradiation time and neutron flux, the activity will increase in direct proportion to the foil thickness, so the thicker the foil the greater will be the exposure of the film for any total neutron dose to the foil. Unfortunately these simple theoretical concepts only give a limited explanation of the observed phenomenaa, and some of the reasons are as follows.

Taking dysprosium and indium as examples we can say that as the foil thickness is increased it approaches the maximum range of the β decay radiation and so it becomes more difficult for this β radiation to escape from the foil. Conversely as the escape path gets longer (i.e. the foil gets thicker) the particle loses energy and is more likely to be at the most effective energy for film blackening ($\sim$ 100 keV [Ref. 27] for beta particles). So the optimum foil thickness for film blackening is dependent upon the range and energy of the emitted particles.

Berger [Ref. 23] gave a constant neutron exposure to converter foils of varying thickness placed in front and behind the film (direct technique). These results were plotted and he found that initially the density increased as the foil thickness increased, but then in levelled out or fell. The optimum speed combination was taken to be the foil thickness and combination (back, front etc.) that gave the highest density. The results, expressed as relative converter foil speeds, are given in Table 1.13.

Because the highest speeds are given by the double converter foil technique, Berger assumed that single foils would be used where improved resolution was required. Thus the relative speed given for single foils in Table 1.13 is based on a compromise between speed and resolution. Berger's resolution data is given in Table 1.14.

It should be noted that the relative speeds of the direct and transfer techniques given in Table 1.13 cannot be compared as there is insufficient data in Berger's paper to make a reliable comparison.

For the transfer data the foils were exposed to give a constant film density after a 3 half-life transfer time. The effects of varying the irradiation and transfer time is discussed later in this section.

All of the data in Table 1.13 was obtained with a monochromatic thermal neutron beam, but Berger states that this shows a reasonable correlation with similar data taken from a reactor beam containing significant intensities of neutrons outside the thermal energy region.

Table 1.13 Relative Speed of Converter Foils 1)

Converter Foil	Technique	Foil Thickness μm Front	Back	Relative Speed
Rh/Gd	Direct	250	50	5.3
Rh/Rh	Direct	250	250	4.7
Gd/Gd	Direct	25	50	3.7
In/In	Direct	500	750	3.7
Dy/Dy	Direct	150	250	3.7
Cd/Cd	Direct	250	500	3.3
Ag/Ag	Direct	450	450	2.7
Dy	Direct	–	250	2.5
Gd	Direct	–	25	2.4
Cd	Direct	–	250	2.2
Rh	Direct	–	250	2.1
In	Direct	500	–	1.7
Au	Direct	375	–	1.2
Au/Au	Direct	150	250	1.0
Dy	Transfer	250	–	16.4
In	Transfer	50	–	11.2
Au	Transfer	75	–	1

1) The relative speed for the direct and transfer methods are not comparable.

Table 1.14 Resolution Characteristics of Converter Foils

Converter Foil	Technique	Foil Thickness (μm) Front	Back	Resolution * (μm)	Exposure for AA Film D=1.5 n cm^{-2}
Gd	Direct	–	12.5	30	2×10^8
Cd	Direct	–	75	30	2.7×10^8
Rd	Direct	–	75	50	3.1×10^8
In	Direct	125	–	50	4.1×10^8
Dy	Direct	–	250	50	1.4×10^8
Ag	Direct	125	–	50 - 90	3.9×10^8
Rh/Rh	Direct	250	250	50 - 90	7.7×10^7
In/In	Direct	500	750	90	10^8
Ag/Ag	Direct	450	450	90	1.4×10^8
Cd/Cd	Direct	250	500	500	1.1×10^8
Gd/Gd	Direct	6.25	50	30	8.6×10^7
Rh/Gd	Direct	250	50	30	7×10^7
Au	Transfer	75	–	30	4.3×10^9
In	Transfer	50	–	50	2.9×10^9
Dy	Transfer	250	–	50	2.6×10^8

*) The minimum resolvable separation between 500 μm diameter holes in a 500 μm thick cadmium test piece.

When considering the relative speed of the foils used in the transfer process it must be remembered that the process of exposing a metal foil to a neutron beam and then using the decay activity of the foil to expose a photographic film is conceptually one of indirectly utilising the mass energy of the neutrons via the secondary radiation in order to convert silver halide to metallic silver within the photographic emulsion of the film. The foil may be regarded as a container for the mass energy of the neutron – a container with an exponential profile that fills less and less rapidly as it approaches its state of maximum capacity. This container is then 'emptied into the photographic film', again at an exponential rate, with the rate of transfer getting less and less as the container becomes empty.

The question of exposure and transfer time is therefore one of determining what fraction of the maximum foil activity is to be induced onto the foil and what fraction of this activity is to be transferred to the film. The product of these two fractions will determine the total fraction that is transferred, and thus the total exposure.

Figure 1.19 shows that when indium and dysprosium foils of the same thickness are irradiated in a thermal-neutron beam then:
(a) when irradiated to saturation and transferred to infinity (i.e. $>$ 3 half-lives in each case) the dysprosium is about 5 times faster than the indium
(b) when irradiated for up to 3 hours and transferred to infinity the dysprosium is 2-4 times faster than indium.
But when the irradiation and transfer times are equal it can be shown that because of the more rapid energy transfer during decay, the indium is faster for exposures below about 0.5 hours, after which the dysprosium is progressively faster, rising to about x3 at about a 3-hour exposure and transfer time.
A method of determining the exposure and transfer times is given in Appendix 1.3.

1.2.3.3 Film and Foil Resolution

Berger [Ref. 29] also studied the resolution capabilities of foils by judging the smallest observable space between closely pitched holes, and the values given in Table 1.14 are those foil thicknesses below which this method shows little or no gain in resolution. The resolution test piece used was 0.05 cm thick cadmium plate with a line of 0.05 cm diameter holes at varying separation.

The use of this data to make comparisons between the resolutions effectiveness of various foils should be made with care because resolution is dependent upon contrast and object size [Ref. 44] and the hole-spacing method is a practical way of defining resolution rather than an absolute method. It should be noted that Berger [Ref. 29] also used a gadolinium test piece of 55 μm thickness with which a hole spacing of 10 μm was resolved by a 12.5 μm thick gadolinium converter foil.

Berger also tested the dependance of the results on film grain size and concluded that it was not influencing the results obtained.

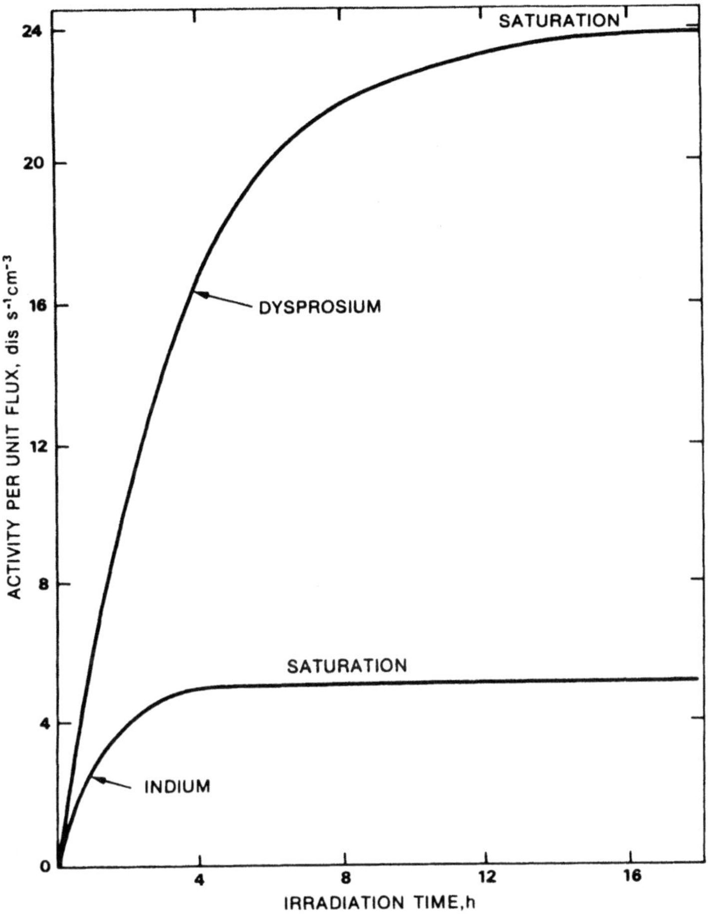

Fig. 1.19 Build-Up of Activity in Indium and Dysprosium of the same Thickness.

1.2.3.4 The Mounting of Converter Foils

Most converter foils used in neutron radiography are between 0.0025 cm and 0.05 cm in thickness, and are, typically, of the order of 200 to 600 cm^2 in area. It is important that these foils remain flat and undamaged so that good contact between the film and the foil is achieved over the whole surface of the foil. The thicker foils will withstand normal handling, but, whilst methods of foil stiffening should be avoided if possible, when very thin metal foils are used they will require the support of a backing plate in order to withstand day-to-day handling. Experience with indium and dysprosium indicates that these foils need to be about 0.08 cm and 0.012 cm thick respectively for use without such backing.

Below these thicknesses the foils should be attached to 0.15 cm thick aluminium plate of at least 99.5% purity. The adhesive used should be as thin as possible as it will usually contain hydrogen, which will scatter the neutrons. A suitable adhesive is photographic mounting tissue. This is applied with a hot iron and so produces a flat, wrinklefree, mounting. For details of the mounting technique see the manufacturer's literature.

1.2.3.5 Enrichment of Converter Foils

For high resolution and short exposure times the higher the foil activity the better will be the results. One possible means of effectively increasing this activity is to make the foil more sensitive to the neutrons by enriching those isotopes which are the most effective for absorbing the neutrons and converting them to film-blackening radiation.

Unfortunately the enrichment process requires special equipment and is likely to be expensive, but it has been reported [Ref. 31] that boron, dysprosium and gadolinium foils have been enriched.

1.3 APPLICATIONS OF NEUTRON RADIOGRAPHY

The listing of reports which describe the applications found up to 1977 has been admirably carried out by John Barton in his edited and indexed compilation of Neutron Radiography Nowsletters, Numbers 1-15, (available form the American Society for Non-Destructive Testing, 3200 Riverside Drive, Columbus, Ohio, 43221), and by the contributers to [Ref. 51]. The following survey gives a general overview of the present situation.

1.3.1 Nuclear Applications

Probably the biggest nuclear use is the examination of experimental fuel pins. The transfer and track-etch methods make such radiography possible, and considerable data can be obtained on cracking, slumping, swelling, etc. By very careful techniques the dimensional changes of the fuel can be measured and then translated into volume changes. The use of this type of application is closely followed by general examination of all types of irradiation experiments for any type of failure that can be detected by visual observation.
A 'marker' technique has been developed whereby the swelling of a pressurised tube in an irradiation experiment (see Figure 1.20) is followed by a pair of plungers which are marked by small washers of dysprosium. This material has a large neutron cross-section and the marker consequently shows on a radiograph as a fine, high contrast line. The distance between two such lines is directly related to the diameter of the tube, and by comparing this with the distance between two other fixed markers of known separation, also within the experiment, the seperation of the measuring markers can be gauged precisely. This method measures the growth of the tube to $\pm$ 25 μm.

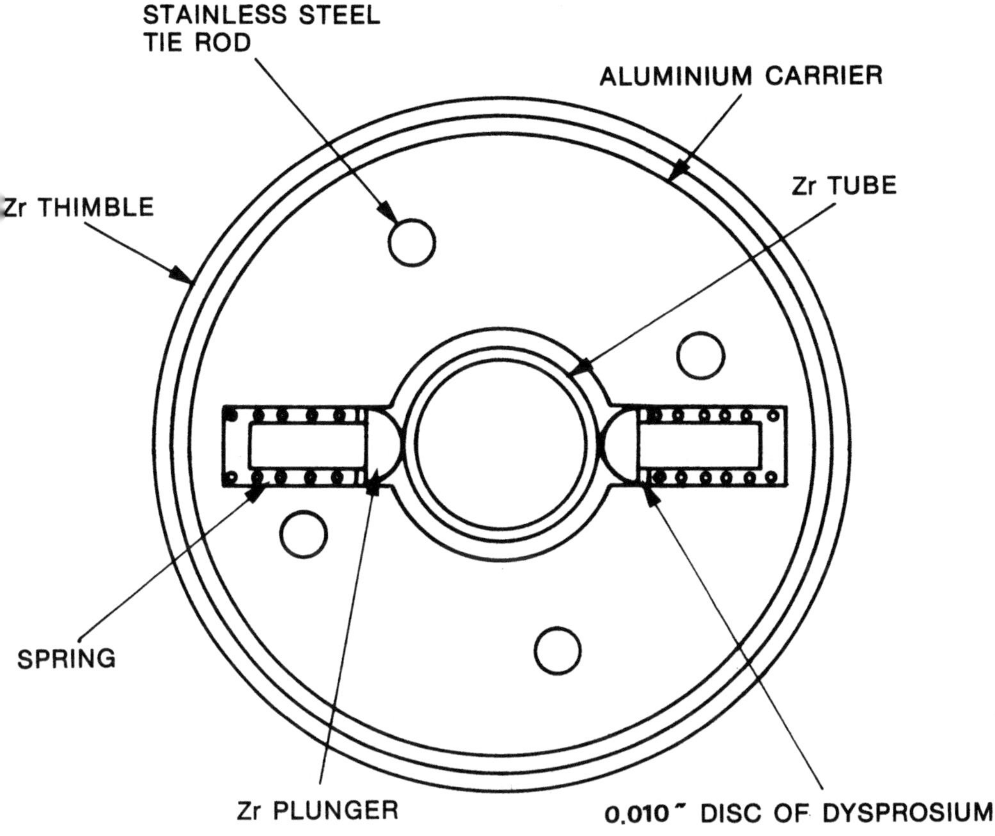

Fig. 1.20 Cross-Section of Pressure-Tube Rig.

Nuclear reactors are controlled by inserting highly neutron-absorbing materials into the pile. As irradiation proceeds the atoms of this material undergo transmutation, causing a marked change in neutron-attenuation cross-section. The rate of depletion of such control materials is of considerable interest to reactor operators as this determines the life of the control absorber and hence its planned replacement. The 'burn-up' of control absorbers can be detected by taking regular neutron radiographs and measuring the size of the depleted areas as is shown by Figure 1.21.

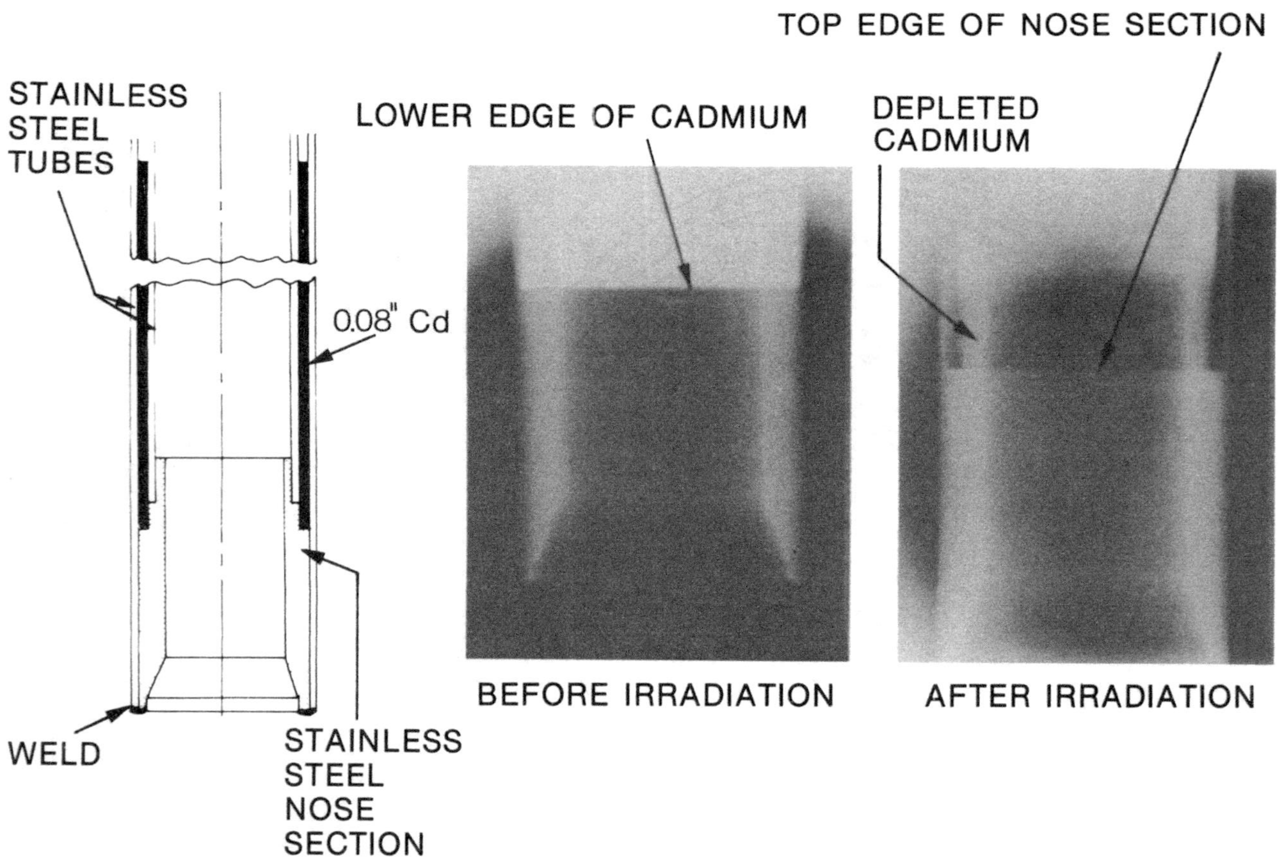

Fig. 1.21 Neutron Radiographs showing Burn-up of Cadmium in Vertical Control Rods.

When neutron shields are built around nuclear installations it is necessary to check their integrity. A typical example is the inspection of a resin filled shield plug where neutron radiography is used to check that the resin has flowed into all the extremities of the volume to be filled.

A neutron radiograph will distinguish between the isotopes of many materials since these often have very different neutron cross-sections. For example U^{235} has a thermal-neutron cross-section of 100.5 barns whilst the cross-section of U^{238} is 2.7 barns. Such differences are readily detectable and allow experimental fuel elements to be checked for rogue fuel pellets.

Neutron radiography has been in use by many workers for the quantitive measurement of hydrides in zironicum-hydride. This is a nuclear problem associated with water reactors in which a corrosion reaction occurs between the water and the zirconium to produce zirconium hydride. The detection technique is non-destructive and provides a two dimensional survey of the hydride concentration in the area under examination. A commercial neutron radiography service offers the detection of hydrogen in zircaloy to a sensivity of 3 ppm-cm.

It must be made clear that this method only detects a material of high neutron attenuation cross section, and it is not able to label an individual element. When detecting hydrogen is zirconium-hydride the experimenter knows that the parent material is pure zirconium and that only high neutron-attenuation cross-section material is present, namely hydrogen.

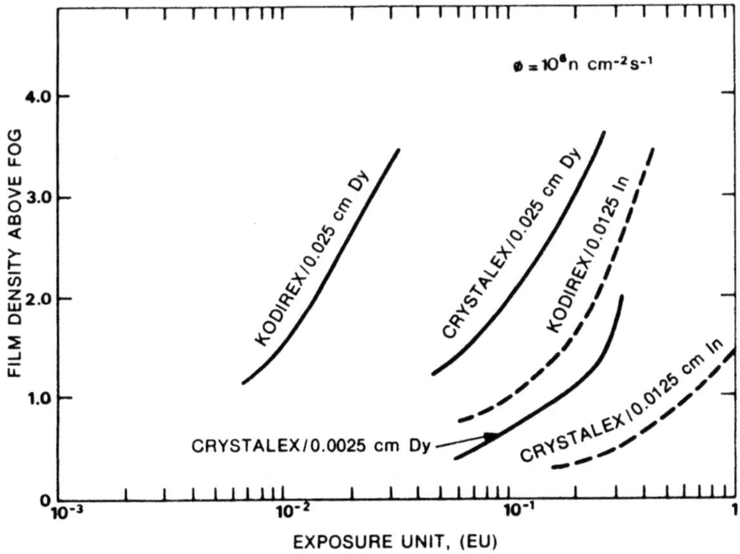

Fig. 1.22 Characteristic Curves for some Film-Foil Combinations.

1.3.2 Industrial Applications

Hydrogen has a large thermal-neutron cross-section and many of the most widely used applications of neutron radiography involve its detection. Rubber and plastic materials have many hydrogen atoms in their molecular structure and so rubber seals, plastic insulation, etc. are easily detected in sealed assemblies. Explosives are also rich in hydrogen and the presence of voids, blockages etc. in ordnance components can be seen (see Figure 1.23). Quantitative measurements of hydrogen have been made to determine absorbed hydrogen in getters and a commercial neutron radiography service offers the detection of hydrogen in zircaloy to a sensivity of 3 ppm.cm.

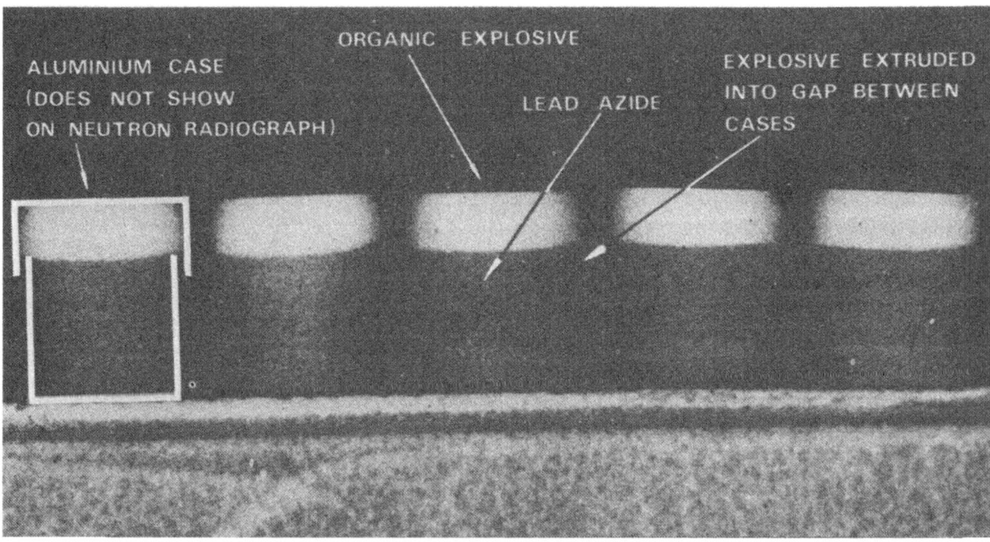

Fig. 1.23 Neutron Radiograph of Explosive Detonators (Magn. ca. 5x)

Brazing and soldering meterials are good subjects. The braze usually contains silver and boron and the flux also contains boron, so both of those can be detected by neutrons. This often makes it possible to detect a dry joint by the presence of excess flux, and the correct flow and penetration of the braze into the joint can be seen from the presence of the boron (see Fig. 1.24).

Turbine blades contain small cooling passages through the length of the blade and neutron radiography has been used to establish the thickness of metal round the passages prior to machining to outer surface of the blade and to identify materials causing blockages in the passages.

Aircraft engine parts have been inspected for the presence of solidified oil and grease in lubrication holes and passages.

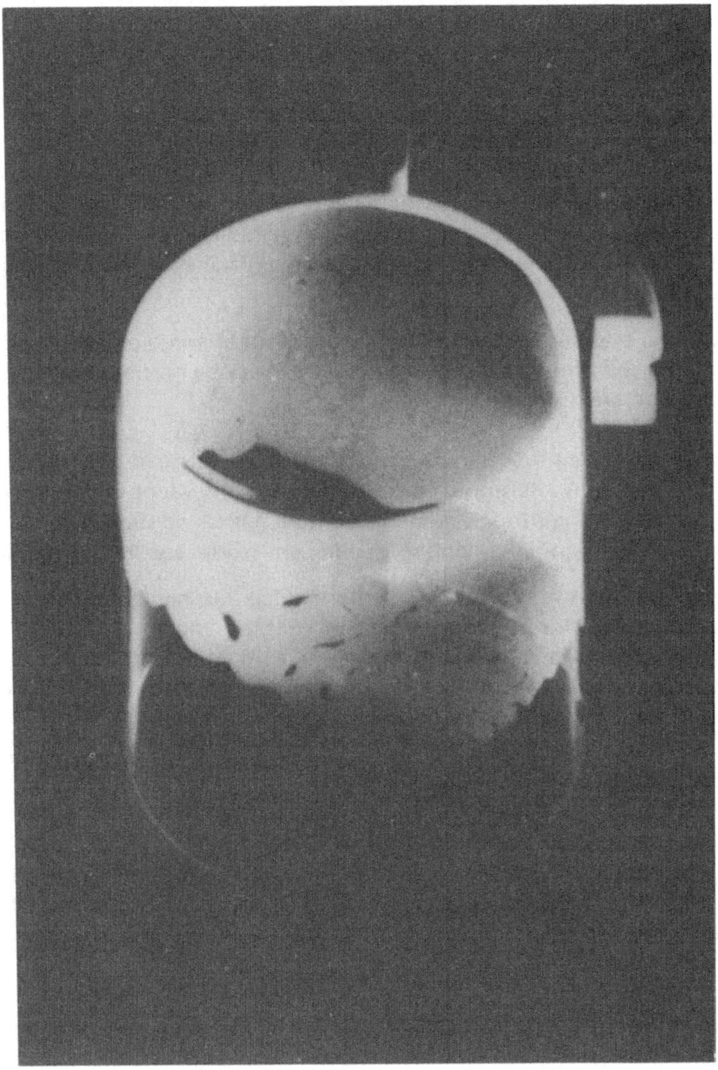

Fig. 1.24 A Brazed Joint between Two Concentric Cylinders.

Racing-car wheels are made of magnesium alloy, and epoxy resins are used in their construction. A combination of ultrasonic methods and neutron radiography has been used to inspect the resin joints.

Helicopter blades have been constructed by bonding carbon fibres to steel, and the lay-up of the fibres has been inspected through the steel by imaging the resin used in the bond.

Printed circuits have been contructed with epoxy resins between layers of copper and neutron radiography has been used to detect voids in the resin.

Soldered joints sometimes exhibit poor electrical characteristics due to contamination within the joint. Several such joints have been neutron radiographed and a contaminant, probably boron, has been detected.

High pressure hose has a metal braid wrapped about a rubber tube and NR has been used to examine the rubber through the metal (steel) braid.

Laminations of various forms are widely used throughout industry and many of these use epoxy resins as the adhesive. Neutron radiography has been used to examine bonded wooden aircraft floors, aluminium honey comb sections for aircraft structures etc.

Friction welding has been used to join stainless steel and aluminium tubes in which the weld is formed at a conical joint. Inclusions and poor bonding have been detected in such joints.

Run-out on deep drilled holes occur, and sometimes this can only be detected by neutron radiography. Run-out has been measured on deep-drilled molybdenum bars using water as a contrast agent in the hole. Other contrast agents which have been used are parafinn, alcohol, gadolinium oxide, and boron fluoride.

Undoubtable the most impressive industrial application has been the cold neutron radiography of a running aircraft gas turbine engine in order to establish the dynamic distribution of the lubricationg oil throughout the oil-passages within the engine. This type of examination is claimed to lead to significant reductions in the time to develop new aircraft engines.

1.3.3 Biomedical Applications

Whilst a number of experiments have been performed in the field of application there have not been any that have shown significant advantage over other methods. This is principally because the neutron has a greater biological effect (to a patient) than photons per unit of absorbed dose and because the required exposures are too high.

Figure 1.26 shows the structure of grass and leaves and is a simple illustration of a biological specimen.

1.3.4 Other Applications

Figure 1.25 shows a cigarette lighter and illustrates how the hydrogen in the petrol is more readily detected by neutron than by X-rays. The flint and the braze metal show-up well and the fibre sealing washer is clearly seen, again due to the hydrogen content.

Perhaps of greater interest are the archaeological applications in which a saxon shield boss was examined and information obtained on the metal joining techniques that had been used. Examination of a Roman spearhead showed that a type of wiped lead joint had been used between the head and the shaft.

X RADIOGRAPH

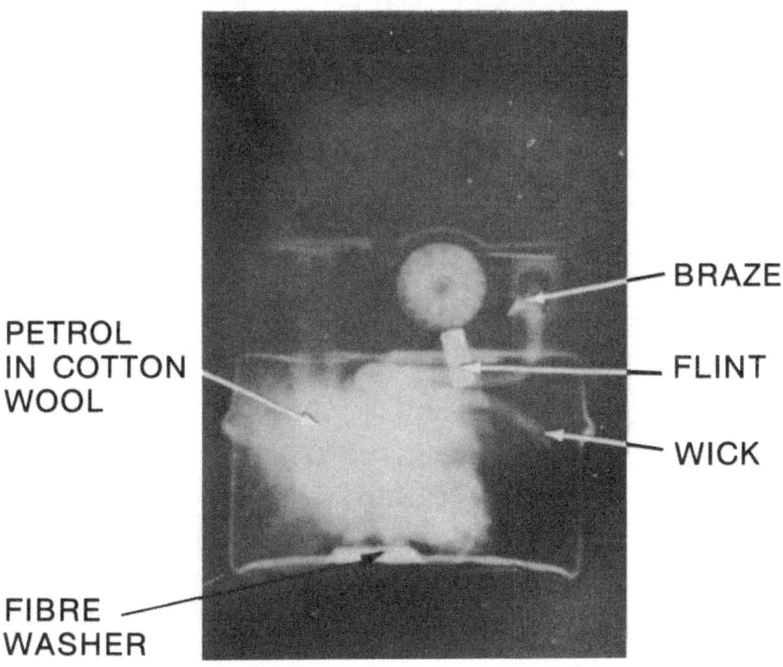

NEUTRON RADIOGRAPH

Fig. 1.25 X-Radiograph and Neutron Radiograph of a Cigarette Lighter.

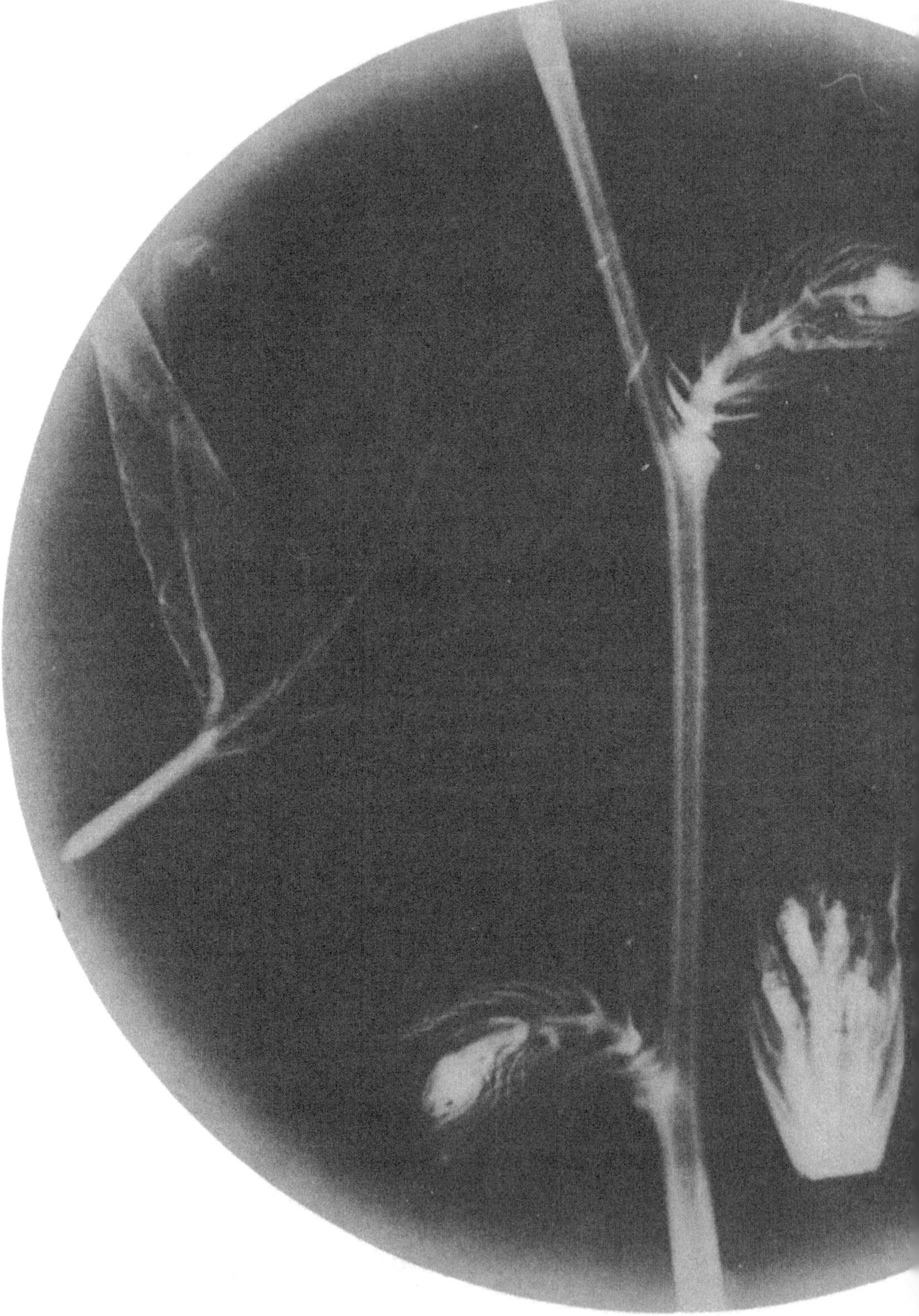

Fig. 1.26 Neutron Radiograph of Grass and Leaves.

1.4 LIST OF SYMBOLS USED

A	attenuation factor
A	atomic weight
A_0	atomic number
B	neutron transmission $= e^{-\Sigma x}$
C	concentration of hydrogen in zirconium, ppm
D	diameter of collimator inlet aperture, cm^2
D_b	background density of film
D_e	density of film
D_i	image density of film
D_{max}	maximum density attainable of film
E	exposure, beta cm^{-2} or neutrons cm^{-2}
E_d	exposure due to direct neutron flux, n cm^{-2}
E_0	interception on film characteristic curve
E_s	exposure due to scattered neutron flux, n cm^{-2}
EU	exposure unit $= 1 - e^{-\lambda T}$
F_c	collimation factor $= \frac{\phi_i / \phi_o}{(L/D)^2}$
F	fraction of particles that emerge from a foil
G	average slope of film characteristic (density/exposure) curve
I	intensity of transmitted light from a film
I_0	intensity of incident light on a film
J	proportionality constant
K	the number of times the density due to the void is greater than the standard deviation in the density of similar areas of sound object
L	length of collimator, cm
L_f	image to object distance, cm
L_s	source to object distance, cm
N	number of atoms or nuclei, in cm^{-3}
N_A	Avogadro's Number $= 6{,}002 \times 10^{23}$
R	range of beta particles in foil material, cm
S	activity, disintegrations.s^{-1}
T	irradiation or exposure time, s
U_i	inherent unsharpness, cm

U_g	geometric unsharpness, cm
U_{ff}	foil and film unsharpness, cm
U_t	total unsharpness, cm
$\triangle D$	minimum detectable or observable density change
$\triangle E$	minimum detectable density change
$\triangle x$	minimum detectable thickness change in the object, cm
a	area of collimator at exit, cm^2
b	distance from film surface to the emulsion, cm
d	foil or filter thickness, cm
h	height of foil, cm
ho	height of image
l	length of collimator wall which emits neutrons, cm
t	transfer time, s
t_1	time interval from end of irradiation to commencement of film exposure, s
t_2	time interval from end of irradiation to completion of film exposure
w	width of specimen, cm
x	specimen thickness, cm
x	minimum detectable thickness variation of void in sample, cm
Σ	macroscopic total neutron cross section, cm^{-1}
Σ_c	macroscopic total neutron cross section of a compound, cm^{-1}
Σ_h	macroscopic total cross section of hydrogen per ppm of hydrogen in zirconium, cm^{-1} ppm^{-1}
Σ_{Zr}	macroscopic total cross section of zirconium, cm^{-1}
δ	see Fig. 1.26
ϵ	Semi-anale subtended to the length at the foil at the point of observation, radians
η	efficiency of the film/foil combination to register neutron events
θ	angle of emission of beta ray from foil, degrees
λ	decay constant $= 0.69/\gamma$, s
μ	film constant, $cm^2 beta^{-1}$
μ_m	mass absorption coefficient, $cm^2 g^{-1}$
ρ	density, g cm^{-3}

ρ_c	film density, $mg.cm^{-3}$
ρ_m	density of material required to stop a beta particle, $mg.cm^{-3}$
σ	microscopic total neutron cross-section, cm^{-2}
τ	half life, s
ϕ_0	neutron flux at exit of collimator, or incident on a sample, $n\ cm^{-2}s^{-1}$
$\bar{\phi}$	neutron emission at exit of collimator, $n\ s^{-1}$,
ϕ	neutron flux transmitted through a sample or incident upon a foil, $n\ cm^{-2}s^{-1}$
ϕ_d	direct neutron flux, $n\ cm^{-2}s^{-1}$
ϕ_i	neutron flux at entrance to a collimator, $n\ cm^{-2}s^{-1}$
ϕ_s	neutron source flux, $n.cm^{-2}s^{-1}$
ϕ_r	radiation flux, particles $cm^{-2}s^{-1}$
ϕ_t	transmitted neutron flux, $n\ cm^{-2}s^{-1}$
$\partial\phi / \partial z$	flux gradient

1.5 REFERENCES

1. H. Kallman, Research 1 (1947-48), p.254.
2. O. Peter, A.E.R.E. Trans. 1111 (1946).
3. J. Thewlis and R.T.P. Derbyshire, A.E.R.E. M/TN 37 (1956).
4. H. Berger. Neutron Radiography, Methods, Capabilities and Applications. Elsevier Publ. Co. (1965).
5. J.P. Barton. Contrast Sensitivity in Neutron Radiography. Appl. Materials Res. 4, No. 2 (1965), p.90-96.
6. Cyril F. Krolick et al. A Technical and Economic Evaluation of Neutron Sources for Thermal Neutron Radiography. TID 25050 (1968).
7. C.A. Hunt. Neutron Radiography with a 5.5 MeV Linear Accelerator Beryllium Source. RARDE Memo 8/69 (1969).
8. J. Stade, Neutron Radiography with an Electron Linear Accelerator. Materialprüfung 19 (1977) N^{0} 10, pp. 436-439
9. G. Olive, J.F. Camerson and C.G. Clayton. A Review of High Intensity Neutron Sources and their Applications in Industry. AERE-R3920 (1962).
10. L. Holland and M.R. Hawkesworth. Low Voltage Particle Accelerators for Neutron Generation. N.D.T. Oct. 1971, p.330.
11. M.R. Hawkesworth and J. Walker, Review: Radiography with Neutrons. J. Mat. Science 4 (1969), p.817-835.
12. M.R. Hawkesworth and R.S. Matfield. The Use of Van der Graaff Accelerators for Neutron Radiography. AERE-R7194.
13. J.P. Barton. Role of Accelerators in Neutron Radiography.
14. Laderer, Holland and Perlman. Table of Isotopes. (1967).
15. R.S. Matfield. The Creep Measurement of Irradiated Specimens by Neutron Radiography. AERE-R5792.
16. J.P. Barton. Contrast Sensitivity in Neutron Radiography. Appl. Materials Res. April 1965, p.90-96.
17. Crosbie et al. Performance of a Californium Multiplexer (CFX) for Neutron Radiography. Nuclear Radiation Newsletter N^{0} 15, Summer 1977.
18. Rockwell. Reactor Shielding Manual, 1st Ed., 1956.
19. M.R. Anderson and D.H.C. Harris. Engineering Applications of Cold Neutron Radiography. Harwell Refort MPD/NB5/107.
20. M.R. Hawkesworth. Some Measurements of the Resolving Power of the Intensifying Screens Used for Radiography with Beams of Thermal Neutrons. J. of Physics and Science Instr. 3 (1970), p.851-854.
21. R.S. Matfield. Neutrons in Radiography, Industry and Commercial Photo. August 1971.
22. W.L. Whitemore et al. A Flexible Neutron Radiography Facility using a Triga Reactor Source. Materials Evaluation, May 1971, p.93-98.
23. M.R. Hawkesworth. Ph.D. Thesis: Characteristics of the Film/Intensifying Screen Image Recorders used in Neutron Radiography. Oct. 1968.
24. D.J. Hughes and R.B. Schwartz. Neutron Cross-SectionsS BNL 325. Second Edition. Supplement 2.
25. Reactor Physics Constants ANL 5800, p. 631.
26. W. Seelmann-Eggebert et al. Chart of the Nuclides (1974).
27. Cobb and Solomon. The Detection of Beta Radiation by Photographic Film. Rev. Sci. Inst. 19, 7th July 1948.

28. H. Berger. A Summary Report on Neutron Radiography. ANL-6846. (1964).
29. H. Berger. Resolution Study of Photographic Thermal-Neutron Image Recorders. J. Appl. Phys. 34, 4th April 1963.
30 M. Müllner and H. Tex. Converter Thickness and Optimum Intensity in Neutron Radiography. Nucl. Instr. Methods (103) 1972, pp229-233.
31. J. Barton and F.M. Klozar. Experiments with Am^{241} –Cm^{242}–Be for Neutron Radiography. AED Conf. 053-003. (1971).
32. R.S. Matfield et al. The Measurement of Small Dimensional Changes by Neutron Radiography. AERE-R.5792.
33. A.R. Spowart. The Advantages of Epicadmium Neutron Beams in Neutron Radiography. Nondestructive Testing 1, p. 151-55, Feb. 1968.
34. M.L. Mullender and V.J. Hart. Transient Neutron Radiography on the Viper Pulsed Reactor. Conf. Radiography with Neutrons, Birmingham, Sept. 1973.
35. Data Sheet on the T X 1432 Tube Supplied by Thomson-CSF, Boulogne, France.
36. G. Farny. Neutron Radiograpy of Irradiated Fuel Elements Using Cellulose Nitrate Film. Conf. Radiography with Neutrons. Birmingham, Sept. 1973.
37. R. Barbalat. Use of the Neutron Radiography Installation for Radioactive Materials on the ISIS Reactor at CEN, Saclay. D.CEN/SPS/EAR/81/091, May 1981.
38. L.G. Miller and T. Watanabe. Enhancing Contrast of Neutron Radiographs by Energy Tailoring of Beams. Conf. Radiography with Neutrons. Birmingham, Sept. 1973.
39. Jaeger et al.. Engineering Compendium on Radiation Shielding, p. 13-14, IAEA 1965.
40. R. Halmshaw. The Physics of Industrial Radiography, Heywood, p. 198.
41. L.E. Wilson et al.. Industrial Development and Application of the Van der Graaff Accelerator for Neutron Radiography. Mat. Evaluation, April 1971, p. 69-74.
42. R. Halmshaw. Research Techniques in Nondestructive Testing, Vol. 2, Academic Press 1973, pp. 31-64.
43. A.A. Halms a B.K. Garside. Radiographic Image Resolution Based on Edge Spread Function Analysis. Trans. Am. Nucl. Soc., Vol. 18, June 74, pp. 62-63.
44. R. Halmshaw. The Factors Involved in an Assessment of Radiographic Definition. J. Photo. Sci. Vol. 3, p. 161-168, 1955.
45. D.C. Cutforth and V.G. Aquino. Neutron Radiography of Radioactive Objects in Hot Cells. Proc. 15th Conf. on Remote Systems Tech., pp. 99-100, (1967).
46. Kodak polarised filters.
47. J.P. Barton et al. Penetrating Neutron Radiography of Fuel Subassemblies. Am. Nucl. Soc., San Francisco, Winter 1977.
48. H.A. Klasens. Measurement and Calculation of Unsharpness Combinations in X-Ray Photography. Phillips Research Report, 1, 4, (Aug. 1946).
49. Eastman Kodak Co. Physical Properties of Kodak Ester Base Films for Graphic Arts.

50. H.P. Leeflang. An apparatus for Neutron Radiography. RCN-135, Petten, March 1971.
51. Atomic Energy Reviews, Vol. 15, No. 2, June 1977.
52. M. Copic et al. On the Inherent Unsharpness of some Radiographs and Neutrographic Testing Methods of Active Fuel Elements. Materialprüfung, Vol. 18, No. 5, pp 171 - 175, May 1976.
53. Radiation Sources for Laboratory and Industrial Use. Catalogue of Radiochemical Centre Amershaw, U.K. (1976).
54. M.R. Hawkesworth. Neutron Radiography: Equipment and Methods. Atomic Energy Reviews, Vol. 15, No. 2, June 1977.
55. Aerotest Operations Report AO 77-27, June 1977.

APPENDIX 1.1
NEUTRON RADIOGRAPHIC TERMINOLOGY

absorption coefficient:	related to the rate of change in the intensity of beam of radiation as it passes through matter.
absorption cross section:	the probability expressed in barns, that a neutron will be totally absorbed by the atomic nucleus.
activation:	the process of causing a substance to become artificially radioactive by subjecting it to bombardment by neutrons or other particles.
attenuation:	the loss of power suffered by radiation as it passes through matter.
attenuation coefficient:	the interaction probability of neutrons per unit path length (cm^{-1}). It is the same quantity as the macroscopic cross-section.
barn:	unit of area for measuring the cross-section of nuclei (1 barn $= 10^{-24}$ cm^2).
BPI:	Beam Purity Indicator, a device for measuring the composition of the beam used in neutron radiography.
cadmium ratio:	ratio of the activity induced by the neutron beam in a bare gold foil to that induced when the foil is covered with cadmium.
cold neutrons:	see thermal neutrons.
collimator:	device for obtaining a neutron beam of small angular spread.
collimator ratio:	also called L/D-ratio, where L is the collimator length and D is the characteristic entrance diameter of the collimator.
contrast agent:	a material added to a component to enhance details by selective absorption of the incident radiation.
contrast capability:	the smallest inclusion, thickness change or density change that can be perceived on the radiographic film, expressed as a percentage.

conversion screen: also called converter, a material placed in contact with the radiographic film, that absorbs neutrons and emits ionising radiation thereby exposing the film.

cassette: a light-tight device for holding film or conversion screens and film in close contact during exposure.

cross section: the apparent cross sectional area of the nucleus as calculated on the basis of the probability of occurence of a reaction by collision with a particle.
It does not necessarily coincide with the geometrical cross-sectional area πR^2. It is given in units of area (barns).

direct exposure imaging: in the direct exposure imaging method the conversion screen and image recorder are simultaneously exposed to the neutron beam.

direct imaging method: method by which the neutron radiation is recorded immediately after passing through the material being tested.

electron volt: the kinetic energy gained by an electron after passing through a potential difference of one volt.

epithermal neutrons: neutrons which have energies in excess of the energy associated with thermal agitation. Neutrons which have speeds and energies intermediate between fast and thermal neutrons (i.e. between about 0.1 and 100 eV).

filtered neutron beam: neutron beam after passing through a uniform layer of material for the purpose of absorbing specific parts of the neutron spectrum.

flash neutron radiography: technique using a neutron source that yields a very high flux neutron beam during a very short time.

gamma ray: electromagnetic radiation having its origin in an atomic nucleus.

gamma ray fogging: increase in the optical density of a radiograph caused by the gamma radiation emitted by the neutron source, by the facility itself, by the object being tested or by a combination of them.

geometric resolution: smallest theoretical size of discontinuity that can be detected according to the geometry of the neutron radiography facility. It depends on the L/D ratio of the collimator, the neutron source to object distance and the object to converter distance.

half value layer:	the thickness of an absorbing material required to reduce the intensity of a beam of incident radiation to one-half of its original intensity.
imaging method:	method by which the neutron radiation passing through a material is recorded.
image quality indicator:	a device or combination of devices whose image or images give a measurement of the neutron radiographic image quality.
indirect exposure imaging:	in the indirect exposure methode, only a gamma insensitive conversion screen is exposed to the neutron beam. After exposure the conversion screen is placed in contact with the image recorder.
indirect imaging method:	also called activation transfer method. Method by which the neutron radiation passing through the material being tested is used to activate a foil of a suitable material. This activated foil is subsequently placed in contact with a medium capable of recording the radiation emitted as the activity of the foil decays.
in-motion neutron radiography:	neutron radiography on moving objects by means of techniques allowing multiple exposures of short duration.
IQI:	Image Quality Indicator.
L/D ratio:	one measure of the resolution capability of a neutron radiographic system. It is the ratio of the distance from the entrance aperture to the image plane (L) to the diameter of the entrance aperture (D).
linear absorption coefficient:	the fractional decrease in radiation beam intensity per unit of distance (cm^{-1}).
mass absorption coefficient:	the fractional decrease in radiation beam intensity per unit of surface density ($cm^2 . g^{-1}$)
moderator:	a material used to slow down fast neutrons. Neutrons are slowed down when they collide with atoms of light element such as hydrogen deuterium, beryllium and carbon.
neutron:	a neutral elementary particle having an atomic mass of 1. In the free state outside of the nucleus, the neutrons is unstable having a half-life of approximately 12 minutes.

neutron radiography:	a process of making a picture of the internal details of an object by the selective absorption of a neutron beam by the object.
neutron to gamma ratio:	ratio of neutron flux and gamma dose rate at the image plane of a neutron radiography facility ($n.cm^{-2}.mR^{-1}$).
recording medium:	a film or detector that converts radiation into a visible image.
saturation effect:	occurring at the activity transfer technique where the activity of the converter, induced by the neutron radiation, increases exponentially to a saturation value where activation and decay are in equilibrium.
scatter factor:	ratio of scattered and non-scattered neutrons that contribute to the resulting visible image.
scattered neutrons:	neutrons that have undergone a scattering collision but still contribute to the resulting visible image. These neutrons may be facility scattered or object scattered neutrons.
sensitivity indicator:	(SI) = a device for indicating the sensitivity of detail visible on a neutron radiography. It is determined by the smallest observable hole and thickness of the corresponding absorber in the indicator.
sensitivity level:	the level determined by the smallest standard discontinuity in any given sensitivity indicator observable in the radiographic film. Levels are defined by identification of type of indicator, size of defect and the absorber thickness on which the discontinuity is observed.
sub-thermal neutrons:	neutrons having energies below 0.01 eV.
thermal neutrons:	neutrons of very slow speed and consequently of low energy. Their energy is of the same order as the thermal energy of the atoms or molecules of the substance through which they are passing; i.e. about 0,025 electron-volts which is equivalent to an average velocity of about 2200 metres per second. Thermal neutrons are responsible for numerous types of nuclear reactions, including nuclear fission.
total cross section:	the sum of the absorption and scattering cross sections.

track-etch imaging:	method by which neutron radiation passing through the material being tested is used to cause damage tracks in a dielectric medium. The damage tracks are made visble by chemical etching.
vacuum cassette:	a light-tight device having a flexible entrance window which operated under a vacuum, holds the film and conversion screen in intimate contact during exposure.

APPENDIX 1.2 THERMAL CROSS SECTIONS OF THE ELEMENTS AND SOME MATERIALS.

Table: Thermal Neutron Cross Sections of the Elements.

Atomic Number	Element or Materials	Density gm cm^{-3}	Nuclei or Molecules $cm^{-3} \times 10^{24}$	Macroscopic Cross Section, cm^{-1}			Absorption Coefficient $cm^2\ g^{-1}$		
				Absorption	Scattering	Total	Absorption	Scattering	Total
1	H	8.99×10^{-5}	5.37×10^{-5}	1.7×10^{-5}	2×10^{-3}	2×10^{-3}	0,196	23	23.2
2	He	1.78×10^{-4}	2.68×10^{-5}	2×10^{-7}	2.1×10^{-3}	2.1×10^{-5}	0	0.1	0.1
3	Li	0.534	0.046	3.29	0.065	3.85	6,1	0.1	6.2
4	Be	1.84	0.123	1.24×10^{-3}	0.865	0.865	0,0067	0.47	0.48
5	B	2.45	0.136	103	0.549	104.6	42,1	0.2	42.3
6	C	1.60	0.080	2.6×10^{-4}	0.385	0.385	0.00018	0.24	0.24
7	N	1.25×10^{-3}	5.38×10^{-5}	9.9×10^{-4}	5×10^{-4}	6×10^{-4}	0.081	0.43	0.51
8	O	1.43×10^{-3}	5.38×10^{-5}	–	2.1×10^{-4}	2.1×10^{-4}	$<7\times10^{-6}$	0.16	0.16
9	F	1.7×10^{-3}	5.39×10^{-5}	10^{-7}	2×10^{-4}	2×10^{-4}	$<3\times10^{-4}$	0.12	0.12
10	Ne	9.0×10^{-4}	2.69×10^{-5}	2.6×10^{-5}	6.2×10^{-5}	8.9×10^{-5}	0.0009	0.072	0.073
11	Na	0.971	0.025	0.013	0.102	0.115	0.0139	0.10	0.11
12	Mg	1.74	0.043	3×10^{-3}	0.155	0.158	0.0016	0.089	0.091
13	Al	2.7	0.060	1.4×10^{-2}	8.4×10^{-2}	9.8×10^{-2}	0.0053	0.031	0.036
14	Si	2.35	0.050	7×10^{-3}	8.9×10^{-2}	9.6×10^{-2}	0.0034	0.036	0.039
15	F	1.83	0.036	7×10^{-3}	0.177	0.184	0.0039	0.10	0.10
16	S	2.1	0.040	1.9×10^{-2}	4.3×10^{-2}	6.2×10^{-2}	0.0098	0.021	0.031
17	Cl	3.21×10^{-3}	5.45×10^{-5}	2×10^{-3}	8×10^{-4}	2.8×10^{-3}	0.58	0.27	0.85
18	A	1.78×10^{-3}	2.68×10^{-5}	2.6×10^{-5}	3.9×10^{-5}	5.5×10^{-5}	0.0099	0.02	0.03
19	K	0.87	0.013	2.6×10^{-3}	2×10^{-3}	4.7×10^{-3}	0.0318	0.023	0.053
20	Ca	1.54	0.020	10^{-2}	7×10^{-2}	8×10^{-2}	0.0066	0.046	0.053
21	Sc	2.5	0.034	0.787	0.804	1.59	0.32	0.4	0.72
22	Ti	4.5	0.057	0.328	0.226	0.555	0.072	0.05	0.12
23	V	5.96	0.071	0.359	0.352	0.711	0.059	0.059	0.118
24	Cr	6.92	0.080	0.238	0.247	0.485	0.036	0.035	0.071
25	Mn	7.42	0.081	1.04	0.181	1.22	0.141	0.025	0.166
26	Fe	7.86	0.085	0.215	0.933	1.15	0.0282	0.12	0.15
27	Co	8.71	0.089	3.37	0.637	4.00	0.38	0.071	0.45
28	Ni	8.75	0.090	0.42	1.6	2.02	0.047	0.18	0.23
29	Cu	8.94	0.085	0.313	0.611	0.924	0.036	0.068	0.104
30	Zn	7.14	0.066	7×10^{-2}	0.237	0.307	0.0102	0.033	0.043
31	Ga	5.90	0.051	0.142	0.204	0.346	0.0241	0.036	0.060

Table : Thermal Neutron Cross Sections of the Elements (Continued).

Atomic Number	Element or Materials	Density gm cm^{-3}	Nuclei or Molecules $cm^{-3} \times 10^{24}$	Macroscopic Cross Section, cm^{-1}			Absorption Coefficient cm^2 g^{050-1}		
				Absorption	Scattering	Total	Absorption	Scattering	Total
32	Ge	5.46	0.045	0.105	0.134	0.239	0.02030	0.025	0.045
33	As	3.70	0.030	0.189	0.277	0.466	0.035	0.048	0.083
34	**Se**	4.5	0.034	0.431	0.403	0.835	0.089	0.084	0.173
35	Br	3.12	0.024	0.155	0.41	0.296	0.052	0.045	0.097
36	Kr	3.71×10^{-3}	2.67×10^{-3}	7.3×10^{-4}	7×10^{-4}	1.4×10^{-3}	0.22	0.052	0.27
37	Rb	1.53	0.011	8×10^{-3}	0.13	0.138	0.0051	0.039	0.044
38	Sr	2.54	0.018	2×10^{-2}	0.175	0.195	0.0083	0.069	0.077
39	Y	5.51	0.037	4.8×10^{-2}	0.112	0.160	0.0089	0.0203	0.0292
40	Zr	6.44	0.043	8×10^{-3}	0.338	0.346	0.0012	0.053	0.054
41	Nb	8.4	0.055	$6 c 10^{-2}$	0.272	0.333	0.0075	0.032	0.039
42	Mo	10.2	0.064	0.16	0.448	0.608	0.017	0.044	0.061
43	Tc						0.13		
44	Ru	12.1	0.072	0.179	0.436	0.615	0.016	0.035	0.05
45	Rh	12.4	0.073	11	0.36	11.4	0.88	0.029	0.91
46	Pd	12.2	0.069	0.551	0.248	0.799	0.045	0.020	0.065
47	Ag	10.5	0.059	3.63	0.325	3.98	0.349	0.033	0.382
48	Cd	8.65	0.046	154	0.325	154.3	13.5	0.037	13.5
49	In	7.28	0.038	7.26	8.4×10^{-2}	7.34	1.02	0.011	1.03
50	Sn	7.29	0.037	2×10^{-2}	0.132	0.152	0.0031	0.020	0.023
51	Sb	6.22	0.031	0.182	0.142	0.324	0.0282	0.0213	0.0495
52	**Te**	6.02	0.028	0.133	0.148	0.281	0.0221	0.024	0.046
53	I	4.94	0.024	0.157	0.084	0.241	0.034	0.017	0.051
54	Ke	5.85×10^{-3}	2.68×10^{-5}	10^{-3}	1.2×10^{-4}	1.1×10^{-3}	0.34	0.020	0.36
55	Cs	1.87	8.5×10^{-3}	0.246	0.170	0.416	0.128	0.032	0.160
56	Ba	3.5	0.015	1.8×10^{-2}	0.123	0.141	0.0052	0.035	0.040
57	La	6.15	0.027	0.239	0.403	0.642	0.0386	0.040	0.079
58	Ge	6.9	0.030	2.1×10^{-2}	0.262	0.283	0.0031	0.012	0.015
59	Pr	6.48	0.028	0.324	0.116	0.44	0.048	0.017	0.065
60	Nd	6.96	0.029	1.33	0.464	1.79	0.208	0.066	0.274
61	Pm						0.249	–	0.249
62	Sm	7.75	0.032	255	0.155	255.2	23	–	23
63	**Eu**	5.22	0.021	90.5	0.166	90.6	17	0.033	17
64	Cd	7.95	0.031	**18.83**	–	**18.83**	178	–	178
65	Tb	8.33	0.032	1.39	–	1.39	0.18	–	0.18

Table : Thermal Neutron Cross Sections of the Elements (Continued).

Atomic Number	Element or Materials	Density gm cm^{-3}	Nuclei or Molecules $cm^{-3} \times 10^{24}$	Macroscopic Cross Section, cm^{-1} Absorption	Scattering	Total	Absorption Coefficient $cm^2\ g^{-1}$ Absorption	Scattering	Total
66	Dy	8.56	0.032	34.9	3.17	38.1	3.5	0.37	3.9
67	**Ho**	8.76	0.032	2.05	–	2.05	0.23	–	0.23
68	Er	4.77	0.017	5.44	0.495	5.98	0.58	–	0.58
69	Tm	9.35	0.033	3.93	–	3.93	0.39	0.025	0.42
70	Yb	7.01	0.024	0.878	0.293	1.17	0.128	0.04	0.17
71	Lu	9.74	0.036	3.62	–	3.62	0.38	–	0.38
72	Hf	13.3	0.045	4.71	0.359	5.07	0.341	0.27	0.61
73	Ta	16.6	0.055	1.18	0.277	1.46	0.070	0.017	0.087
74	W	18.9	0.062	2.21	0.316	1.53	0.063	0.016	0.079
75	Re	29.15	0.095	5.58	0.93	6.51	0.28	0.045	0.33
76	Cs	22.5	0.073	1.05	0.783	1.83	0.049	0.049	0.098
77	Ir	22.4	0.078	30.2		30.2	1.37	–	1.37
78	Pt	21.4	0.066	0.535	0.660	1.19	0.027	0.031	0.058
79	**Au**	19.3	0.060	5.79	0.55	6.34	0.302	0.028	0.330
80	Hg	13.6	0.041	14.7	0.814	15.5	1.12	0.06	1.18
81	**Te**	11.9	0.035	0.115	0.489	0.604	0.010	0.041	0.051
82	Pb	11.1	0.033	6×10^{-3}	0.363	0.369	4.9×10^{-4}	0.032	0.032
83	Bi	9.7	0.028	10^{-3}	0.264	0.265	9.8×10^{-5}	0.026	0.026
84	Po	9.24	0.027						
86	Ru	9.73×10^{-3}	2.64×10^{-5}						
87	Fr								
88	Ra	5	0.013	0.266		0.266	0.053	–	0.053
89	Ac						1.35	–	1.35
90	Th	11.5		0.205	0.366	0.571	0.019	0.032	0.051
91	P	15.4		10.4	–	10.4	0.675	–	0.675
92	U	18.7	0.047	0.364	0.397	0.761	0.0193	0.0209	0.0402
93	Np						0.432	–	0.432
94	Pu	19.74	0.049	57	0.478	57.5	2.593	0.024	2.617
95	Am								
96	Cm								
97	Bk								
98	Cf								
99	E								
100	Fu								

Table: Thermal Neutron Cross Sections of some Materials.

Atomic Number	Element or Materials	Density gm cm^{-3}	Nuclei or Molecules $cm^{-3} \times 10^{24}$	Macroscopic Cross Section, cm^{-1}			Absorption Coefficient $cm^2\ g^{-1}$		
				Absorption	Scattering	Total	Absorption	Scattering	Total
	Beo	2.96	0.071	7.3×10^{-4}	0.501	0.501	2.47×10^{-4}	0.169	0.169
	CO_2	1.98×10^{-3}	2.71×10^{-5}	–	2.4×10^{-4}	2.4×10^{-4}	–	0.121	0.121
	D_2O	1.10	0.033	3.3×10^{-5}	0.449	0.449	3×10^{-5}	0.408	0.408
	Dy_2O_3	7.81	0.013	27.7	2.7	30.4	3.547	0.346	3.893
	Eu_2O_3	7.42	0.013	111	0.383	111.4	14.96	0.052	15.012
	H_2O	0.997	0.033	0.022	3.45	3.47	0.022	3.46	3.482
	LiF	2.29	0.053	3.75	0.282	4.032	1.638	0.123	1.761
	UC	13.63	0.033	0.227	0.491	0.718	0.167	0.36	0.527
	UC_2	11.68	0.027	0.185	0.537	0.717	0.158	0.46	0.618
	UN	14.22	0.035	0.237	0.692	0.929	0.017	0.49	0.507
	UO_2	10.8	0.024	0.169	0.372	0.542	0.157	0.034	0.191
	ZrH	5.61	0.036	0.026	2.50	2.526	0.005	0.446	0.451
Aluminium (99,5%)		2.71		0.015	0.084	0.099	5.5×10^{-3}	0.031	0.036
Butyl Rubber		0.92		0.01	1.45	1.46	0.01	1.58	1.59
Concrete (Barytes)		3.3		0.02	0.1	0.2	6×10^{-3}	0.03	0.036
Concrete (Standard)		2.33		6×10^{-3}	0.14	0.15	2.6×10^{-3}	0.06	0.064
Inconel		8.43		0.366	1.225	1.591	0.04	0.15	0.19
Lead ag de		4.8				0.8			0.17
Monel		8.83		0.398	1.258	1.656	0.045	0.142	0.187
Nylon		1.11		0.04	2.56	2.6	0.036	2.31	2.34
Oil		0.88		0.03	3.07	3.1	0.034	3.49	3.52
Paraffin		0.9		0.05	3.1	3.15	0.056	3.44	3.5
Perspex		1.18				2.8			2.37
Polythene		0.94				4.0			1.25
Polystytene		1.07		0.02	2.1	2.12	0.019	1.96	1.98
Rex explosive		1.82				2.9			1.59
Stainless Steel 316		7.92		0.27	0.85	1.12	0.034	0.107	0.14
Teflon		2.17		10^{-3}	0.3	0.3	4.6×10^{-4}	0.138	0.138
Water		1.0				3.45			3.45

APPENDIX 1.3

Irradiation and Transfer Times for the Indirect Method

The formation of a neutron radiographic image on a film, using the transfer method will depend upon:

a) the neutron flux
b) the rate at which the foil material becomes activated and its saturated activity
c) the rate at which this activity decays
d) the fraction of the activation particles that escape from the foil, and
e) the sensitivity of the film to these activation particles.

Assuming that a beta-emitting foil is being used, then these factors may be expressed in an equation for the exposure (E) of the film, as follows [Ref. 32] :

$$\Sigma = 0{,}45 \frac{N \sigma \phi}{\lambda} (1 - e^{-\lambda T}) (e^{-\lambda t_1} - e^{-\lambda t_2}) \cdot \frac{\rho_c}{\rho_m} \cdot d \left(1 - \frac{d}{2R}\right)^{1/2} \text{ particles cm}^{-2}, \qquad (1)$$

where

N = number density of stable isotopes, atoms cm^{-3}
σ = microscopic cross-section, cm^2
λ = $0{,}69/\tau$, where τ is the half-life of foil material, s
ϕ = thermal neutron flux, n cm^{-2} s^{-1}

T = exposure time, s
t_1 = interval from end of irradiation time to commencement of film exposure, s
t_2 = interval from end of irradiation to completion of film exposure, s

ρ_c = film density, mg cm^{-2}
ρ_m = density of material required to stop beta particles, mg cm^{-2}
d = foil thickness, cm (assuming $d < R$)
R = range of betas in foil material, cm.

The constants in this equation are the foil material, the foil thickness, and the speed of the photographic film.
The variable factors are:

$$\phi\, (1 - e^{-\lambda T})\, (e^{-\lambda t_1} - e^{-\lambda t_2}), \qquad (2)$$

i.e. those involving (a) the flux (ϕ)
(b) the irradiation time (T)
(c) the transfer time (t).

Expression(2) is called the integrated exposure, and is the product of the neutron flux, the irradiation unit, and the transfer unit. The product of the last two terms (the quantities in the brackets) is called the exposure unit, or EU, and is the fraction of the maximum possible integrated exposure that is achieved during a particular combination of irradiation and transfer times. Fig. 1.22 shows the characteristic curves for several X-ray films, in which the density above fog* is plotted against the exposure unit (EU) for a neutron flux of 10^6 n cm^{-2} s^{-1}.

*) Density above fog = $\log_{10} I_o/I$, where for a beam of light falling on a film I_o = incident intensity and I = transmitted intensity. Fog is extraneous density or 'noise'.

The required film density can be determined from a knowledge of the image density required and the neutron attenuation through the sample. This is described by the transmission equation

$$B = e^{-\Sigma x}, \qquad (3)$$

where

B = fraction of neutrons transmitted
Σ = macroscopic cross-section of sample, cm^{-1}
x = sample thickness, cm.

APPENDIX 1.4

Resolution of Voids

The effect of the collimator ratio and the neutron exposure on the resolution of voids has been considered by Hawkesworth et al in [Ref. 12] and the following discussion is based on this work.
To detect a void of cylindrical form two requirements must be fulfilled, i.e.

(a) The difference in density between the image of the defect and the surrounding film density must be observable.

(b) There must be a sufficient number of events recorded by the film to ensure that the density of the void can be clearly distinguished from the density variations due to 'noise' in the region of the film surrounding the image of the void.
The first of these requirements can be expressed as

$$\bar{x} = \frac{4\,(\Delta D_e)}{D_e \cdot \Sigma} \qquad (1)$$

where

x = minimum detectable thickness variation or void in the sample, cm
ΔD_e = minimum observable density change
D_e = film density surrounding the image of the object
Σ = macroscopic cross-section of the object, cm^{-1}.

Now for a particular film-foil combination the density may be expressed in terms of the exposure time, i.e.

$$D_e = GT, \qquad (2)$$

where

G = slope of film characteristic (density/exposure) curve
T = irradiation time, s (assuming the foil is transferred to infinity).

Thus

$$\bar{x} = \frac{4\,(\Delta D_e)}{G.T.\Sigma} \qquad (3)$$

The second consideration can be expressed by

$$E = \frac{4K^2}{\pi \cdot \eta \cdot \Sigma^2} \cdot \frac{(U_t + x)^2}{\bar{x}^6} \qquad (4)$$

where

K = the number of times the density due to the void is greater than the standard deviation in the density of simular areas of a sound object
E = exposure at the image, neutrons cm^{-2}
η = efficiency of the film-foil combination to register neutron events.
U_t = total unsharpness, cm
Σ, x = as above.

Now the total unsharpness is a function of the geometric unsharpness and the inherent unsharpness according to

$$U_t = (U_g^{3} + U_i^{3})^{1/3}, \tag{5}$$

where

U_g = geometric unsharpness, cm
U_i = inherent film-foil unsharpness, cm.

But the geometric unsharpness is related to the collimator dimensions by

$$U_g = L_f\left(\frac{D}{L_s}\right) \tag{6}$$

where

L_f = image to obeject distance, cm
L_s = source-to-object distance, cm
D = characteristic dimension of collimator aperture, cm

So from (4), (5) and (6)

$$E = \frac{4K^2}{\pi \Sigma^2 x^6} \{[L_f^{3}(\tfrac{D}{L})^3 + U_i^{3}]^{1/3} + x\}^2. \tag{7}$$

Expression (7) gives the exposure E in neutrons cm^{-2} needed to resolve a defect of characteristic dimension x. This can be related to the attenuated neutron flux at the foil by:

$$\phi = \frac{E}{T}, \tag{8}$$

where

ϕ = attenuated neutron flux at the foil, n $cm^{-2}s^{-1}$
T = irradiation time, s.

Now the unattenuated neutron flux at the foil is related to the collimator dimensions by (Ref. 12):

$$\phi_o = \frac{\phi_i}{16}\left(\frac{D}{L}\right)^2\left(1 + \frac{2l}{L}\right) \tag{9}$$

where

ϕ_i = neutron flux at entrance to collimator, n $cm^{-2}s^{-1}$
ϕ_o = neutron flux at the exit to the collimator, n cm^{-2} s^{-1}
D = collimator aperture, cm
L = collimator length, cm
l = length of collimator wall which emits neutrons, cm.

If we let l = 2D* then

$$\phi_i = \frac{\phi_o}{16}\left[\left(\frac{D}{L}\right)^2 + 4\left(\frac{D}{L}\right)^3\right] \tag{10}$$

The neutrons transmitted through the sample ϕ will be $\phi_o B$ where B is the fraction of the neutrons that are transmitted. Combining (7), (8) and (10) we have:

$$\bar{x}^6 = \frac{64K^2}{\pi \cdot \eta\, \Sigma^2 \cdot T\phi_o \cdot B} \frac{\{[L_f^3(\frac{D}{L})^3 + U_i^3]^{\frac{1}{3}} + x\}^2}{(\frac{D}{L})^2 + 4(\frac{D}{L})^3} \tag{11}$$

*) This will depend on the design of the collimator with respect to the neutron source.

Equations (3) and (11) indicate that the resolution ($\bar{x}$) of a defect is inversely dependent upon the irradiation time (T) and a number of quantities which are characteristics of the collimator, foil, film and sample. These characteristics can be given particular values and the two equations can be solved by simultaneously plotting $\bar{x}$ against T, thereby providing a means of predicting the L/D ratio, exposure, film type, etc. which must be used to detect a particular defect size in a sample. Let us now consider the practical significance of equations (3) and (11).

From equation (3) we have symbol ΔD, which is the minimum density change that can be observed. As this is a characteristic of the given observer and the given viewing equipment it can be regarded as a constant. G is the slope of the film characteristic (density/exposure) curve at the image density which surrounds the defect.

The remaining symbols appear in equation (11). Symbol K expresses the minimum signal-to-noise ratio at which the smallest defect size can be observed. This is a characteristic of the observer and can be regarded as a constant. Symbol η represents the efficiency with which a neutron gives rise to a secondary radiation which activates a clump of silver atoms in the film emulsion. This is a function of the neutron energy, the foil material, the foil thickness, and the film. The need for high efficiency leads to the use of neutron energies which will penetrate the object and yet have the highest possible cross-section for the foil material. The foil should be thin, and earlier discussion on foil thickness will indicate the best to use. Fast films have the best recording efficiency, but this conflicts with the requirement for small grain size.

The symbol Σ is a characteristic of the sample and is therefore not a variable. The symbols T, ϕ_i and B are the irradiation time, the neutron flux at the entrance to the collimator and the transmission of the sample, and these can be considered together as they jointly determine the exposure. For the detection of small defects the exposure should be as large as possible.
Symbol L_f is dependent upon the size of the sample and is therefore not a variable. Symbol D/L is the reciprocal of the collimator ratio; the larger this value the smaller is the detectable defect size.
Symbol U_i is the inherent unsharpness in the foil and film combination, and this should be made as small as possible by using the optimum foil thickness, fine grain film, and good film-foil contact. Good contact between the foil and the film is partly determined by the quality of the foil surface, but mostly by the effectiveness of the cassette used to hold the film and foil in contact. The best type of cassettes used in neutron radiography maintains good film-foil contact by using a vacuum. The use of vacuum cassettes is strongly recommended.
From the examination of the symbols used in equations (3) and (11) let us turn to the values to be used for these symbols. Equation (3) states

$$x = \frac{4(\Delta D)}{GT\Sigma}.$$

where

$\Delta D = 0{,}015$ for observation under good viewing conditions or for when a microdensitometer is used.

The value for G can be obtained from film characteristic curves by assuming that the bulk image density will be between 1,0 and 1,5, and taking the mean value for G between these two densities for films which are appropriate (in terms of speed and resolution) for the work to be performed.
The value to be used fo Σ will depend upon the material(s) of the sample and the neutron energy. Values for thermal neutrons are given in Appendix 1.2.

The remaining symbols are found in equation (11), which stated

$$\bar{x}^6 = \frac{64K^2}{\pi \cdot \eta\, \Sigma^2 \cdot T\phi_o \cdot B} \; \frac{\left\{\left[L_f^3\left(\frac{D}{L}\right)^3 + U_i^3\right]^{\frac{1}{3}} + x\right\}^2}{\left(\frac{D}{L}\right)^2 + 4\left(\frac{D}{L}\right)^3}$$

Hawkesworth et al [Ref. 12] give the value for K as 10, and this is supported by a value of 6-10 derived from a similar measurement by Halmshaw [Ref. 40], so from these we will take a mean value of

$$K = 8$$

We also have $n = 0{,}1$ from [Ref. 12].

The value for ϕ_i is usually known for reactor sources and some values have been published for particular accelerator [Refs. 12, 41] and isotope source [Ref. 45].

The transmission through the sample can be calculated from

$$B = e^{-\Sigma x}$$

where B = fraction of neutrons transmitted
Σ = macroscopic cross-section of sample, cm^{-1}
x = sample thickness, cm

The dimension L_f is usually taken as the distance from the foil to the centre, or front surface, of the specimen, unless there is a particular plane through the specimen that is more appropriate.

The D/L ratio of the collimator will be known if an existing facility is being used. If equations (3) and (11) are being used as the basis for the design of a new facility then several values of D/L can be tried in order to determine the sensitivity of $\bar{x}$ to this term.

The value of U_i, the inherent unsharpness, varies with the film/foil combination and Copic [Ref. 52] has reported values of between 350-670 μm for Dy and In foils (transfer method) with various films. Conversely the values given in Table 1.14 for the total unsharpness of Dy and In with AA film are considerably smaller than the data given by Copic for inherent unsharpness alone. This indicates that the inherent unsharpness must be measured on a representative sample by the method that will ultimately be used to measure the resolution obtained when radiographing the real object.

It is interesting to note that Copic also reported $U_i = 35\ \mu m$ for a $^{10}B\ ^{6}L$ converter and Kodak CA 80-15 type B nitrocellulose film, and that Leeflang and Ruyter measured 28 - 32 μm for the same film/foil combination.

If it is not possible to obtain a realistic value for U_i then values between, say, 35 and 350 should be used to test the sensitivity of the equations to variations in this term.

APPENDIX 1.5

Calculation of the Cross Section of Compounds

Assuming that the property of the nuclear species is unaffected by considerations of the molecular or crystal structure in which it resides (this assumption is usually acceptable for scoping calculation of the type below) then the macroscopic cross section for the compound can be calculated from the summation of the macroscopic cross sections of each nuclear species:

$$\Sigma_c = \Sigma N_i \cdot \sigma_i \qquad (1)$$

where Σ_c = macroscopic cross section of compound, cm^{-1}
N_i = number of nuclei per unit volume in i th nuclear species, nu cm^{-3}
σ_i = microscopic cross section for i th nuclear species, cm^2

Now

$$N = \frac{\rho N_A}{A} \qquad (2)$$

so

$$\Sigma_c = \Sigma\left(\frac{\rho N_A}{A}\right)_i \sigma_i \qquad (3)$$

or

$$\Sigma_c = \rho . \frac{N_A}{A} \Sigma \sigma_i \qquad (4)$$

where

Σ_c = macroscopic cross section of compound, cm^{-1}
ρ = density of compound, g cm^{-3}
N_A = Avogadro's number
A = atomic weight of compound
σ = microscopic cross section of each atom, cm^{-2}

Example of cellulose acetate

ρ = 1,3 g cm^{-3}
N_A = 6,023 x 10^{23}
A = 246,22

The chemical formula for cellulose acetate is

$C_6H_7O_2(OH)(C_2H_3O_2)_2$ i.e.:

10 carbon atoms at 4,8 barns/atom
7 oxygen atoms at 4,2 barns/atom
14 hydrogen atoms at 38 barns/atom

Total cross-section $= 48 + 29{,}4 + 522 = 599{,}4$ barns, say $= 600$ barns per molecule.

Thus

$$\Sigma_c = \frac{600 \times 10^{-24} \times 6{,}02 \times 10^{23} \times 1{,}3}{246{,}22}$$

$$\underline{\Sigma_c = 1{,}9 \text{ cm}^{-1}}$$

2. RECOMMENDED PRACTICE FOR THE NEUTRON RADIOGRAPHY OF NUCLEAR FUEL

a) This part of the Neutron Radiography Handbook is a guide for the satisfactory neutron radiographic testing of nuclear fuel. It relates to the use of photographic film, radiographic film and tracketch recording materials.

b) It includes statements about prefered practice but does not discuss the technical background which justifies the preference. Such background information is given in Part 1 of the Handbook.

c) This document does not recommend a prefered design for the equipment which produces the neutron radiographic beam, or the prefered quality of the beam (neutron energy, gamma contamination etc.). For this data reference should be made to the neutron radiographic principles discussed in Part 1 of this Handbook.

d) This document describes methods of measuring radiographic quality and refers to reference radiographs for nuclear fuel, but it does not cover the interpretation or acceptance standards to be applied as this is considered to be a subject that should be covered by the Order Specification and therefore a matter of contractual agreement between the supplier and the purchaser.

e) The numerical data quoted herein has been taken from Part 1 of the Handbook, which gives the relevants source references.

f) Sections 2.7, 2.8, 2.9, 2.11 and 2.12 of this Recommended Practices have been taken verbatim from ASTM E94-77 'Standard Recommended Practice for Radiographic Testing' and the compilers of this Handbook make grateful acknowledgement to the American Society for Testing Materials for their permission to do this.[1)]

2.1 APPLICABLE DOCUMENTS

a) Neutron Radiography Handbook – Part 1, Principles and Practice of Neutron Radiography.

b) Neutron Radiography Handbook – Part 3, Beam and Image Quality Indicators for Neutron Radiography.

c) Neutron Radiography Handbook – Part 4, Reference Radiographs of Defects in Nuclear Fuel.

d) Neutron Radiography Handbook – Part 5, List of Neutron Radiography Facilities in the European Community.

1) Reprinted with permission from the Annual Book of ASTM Standards, Part 11. Copyright: American Society Society for Testing and Materials, 1916 Race Street, Philadelphia, PA 19103.

2.2 ORDERING INFORMATION

The following list gives the information which is recommended for inclusion in a Purchase Order for the services covered in this recommended practice.

a) Clients name and address.

b) Description of the object to be radiographed.

c) Objective of the neutron radiographic examination, giving qualitative and quantitative information.

d) Information on previous radiographic examinations (including X-radiography, gamma-radiography, etc.).

e) Any radiographic parameters that must be met.

f) Identification requirements.

g) Radiographic density requirements.

h) Radiographic quality as defined by an image quality indicator.

i) Requirements for the written report.

2.3 EQUIPMENT

2.3.1 General

2.3.1.1 Where possible a neutron radiography facility which is most suitable for carrying out the required detection or measurement should be used.
To obtain this requirement the advantages of optimising the geometry, neutron energy, and beam quality should be considered whenever the facility allows these parameters to be controlled.

2.3.1.2 The use of the track etch technique is discussed in para. 2.4.12 and all references to 'film' in the following paragraphs relate to photographic film. Information on track-etch materials is included in the Table 2.5.

2.3.2 Geometry

The geometry may be controlled by varying the size of the beam inlet-aperture, by changing the inlet-aperture to object distance or by changing the object to film distance (see para. 2.4.7). It is recommended that the equipment should have the facility to vary the geometry.

2.3.3 Neutron Energy

2.3.3.1 The control of neutron energy is a function of both the choice of neutron source and the selection of a prefered energy from the available radiation energies in the beam. The first parameter is fixed by the choise of neutron source, as shown in Tables 2.1 to 2.3. The second is controlled by the use of neutron beam filters, and some of these are listed in Table 2.4 (see Part 1 for more information on filters).

TABLE 2.1 TYPICAL ACCELERATOR SOURCE-REACTIONS

Incident particle	Target material	Neutron output energy
Deuteron	Tritium	14 MeV
Deuteron	Beryllium	4 MeV
Deuteron	Lithium	2 MeV
X-rays	Beryllium	1.6 Mev
X-rays	Uranium	
Electrons	Tantalum	1-4 Mev

TABLE 2.2 RADIOISOTOPIC SOURCES

Source	Half-life	Reaction	Neutron yield ($n.s^{-1}\ g^{-1}$)	Neutron Energy (MeV)
^{124}Sb-Be	60 d	(γ,n)	2.7×10^{9}	0.024
^{210}Po-Be	138 d	(α,n)	1.28×10^{10}	4.3
^{241}Am-Be	458 a	(α,n)	1×10^{7}	~ 4
^{226}Ra-Be	1620 a	(α,n)	1.3×10^{7}	~ 4
^{227}Ac-Be	21.8 a	(α,n)	1.1×10^{9}	~ 4
^{228}Th-Be	1.91 a	(α,n)	1.7×10^{10}	~ 4
^{252}Cf	2.65 a	fission	2.34×10^{12}	2.3

TABLE 2.3 NUCLEAR REACTOR SOURCE

Reaction	Neutron output energy
Fission	Predominant 0,03 eV
	Maximum 2 MeV

2.3.3.2 For the neutron radiography of nuclear fuel a beam with a cadmium ratio of at least 0.1 is recommended. It is also recommended that the equipment should be capable of using a cadmium filter to allow radiography with epicadmium neutrons (energy $>$ 0,4 eV).

TABLE 2.4 SOME NEUTRON FILTERS

Element	Minimum Energy, eV	Absorption Cross Section Barns	Maximum Energy, eV	Absorption Cross Section Barns
Beryllium 4	$5x10^{-3}$	0,45	$7x10^{-3}$	6
Boron 5	$6,5x10^{-4}$	$4,5x10^{3}$	$3x10^{3}$	6
Sodium 11	$5x10^{2}$	3,2	$3x10^{3}$	380
Sulphur 16	$7x10^{4}$	0,45	$1,2x10^{5}$	20
Scandium 21	30	20	$4x10^{3}$	40
Iron 26	$2,6x10^{4}$	0,45	$2,9x10^{4}$	3 (av.)
Rhodium 45	1,3	$5x10^{3}$	20	4,5
Cadmium 48	0,18	$8x10^{3}$	10	4,5
Gadolinium 64	$2x10^{-3}$	$1,4x10^{5}$	1,5	35

2.3.4 Beam Quality

2.3.4.1 The measurement of beam quality defines

a) the fast/thermal neutron ratio, i.e. the cadmium ratio,

b) the gamma ray contamination, i.e. n/γ ratio,

c) the degree of scatter in objects with high scattering cross sections, and

d) the geometric resolution.

2.3.4.2 A knowledge of these factors provide the basis for understanding of the variance in radiographic results and so the measurement of beam quality by the use of the beam quality indicator given at para. 3 is recommended.

2.4 RADIOGRAPHIC TECHNIQUES

2.4.1 General

2.4.1.1 The resolution/detection capability of a neutron radiographic technique increases as:

a) the variation in the specimen thickness is decreased,

b) the scattering cross section of the specimen to the incident radiation in the beam is decreased,

c) the difference between the attenuation coefficient of the volume to be detected and the surrounding material in the object is increased,

d) the sensitivity of the detector to the incident radiation in the beam is increased,

e) the scattering cross section of the recording material to the incident particle or photon coming from the detector is reduced,

f) the grain size of the film is decreased.
The following recommendations are intended to give the best possibility of detecting a discontinuity in a nuclear fuel or to measure fuel rod dimensions.

2.4.2 Set-Up, Marking and Identification

2.4.2.1 The neutron beam should be aligned with the middle of the object under examination and normal to its surface at that point. It is essential that any point on the object can be identified with the corresponding point on the radiograph. To achieve this an unambiguous method of marking the object should be used and cadmium or plastic numerals (or other suitable shapes) should be aligned with the marks on the object.

2.4.2.2 Where it is necessary to identify the edge of a specimen that is near-transparent to the incident beam, such as a thin walled zirconium fuel can, then cadmium or plastic markers should, were possible, be placed against the (curved) surface of the specimen in order to precisely locate its position.

2.4.2.3 When using overlapping radiographs the markers should be placed so as to provide evidence that full coverage has been achieved.

2.4.2.4 Each radiograph should be identified by a unique number so that there is a permanent correlation between the object and the radiograph, and where necessary a sketch should be made of the disposition of the radiographic exposures along the specimen.

2.4.3 Image Converters

2.4.3.1 The material of the converter foils should be chosen to give the maximum detection/resolution efficiency. The neutron cross section of the converter material determines its sensitivity to the incident neutrons and it should therefore be selected to compliment the chosen neutron energy. Part 1 of this Handbook gives details of some of the measurements that have been made on the relative speed and resolution of various image converters.
The commonly used image convertors are:
- Indirect technique, dysprosium
- Direct technique, indium and gadolinium
- Track-etch technique, boron and lithium

2.4.3.2 Converter foils should be as thin as possible commensurate with an adequate nuclear thickness (e.g. cross section times thickness) to give the required image density on the recording film and adequate strength for handling. They should also bee smooth, flat and free from kinks and other surface imperfections.

2.4.3.3 Where necessary the converter foils may be supported by an aluminium backing plate. The prefered method of attachment is by metalic bonding such as spluttering or roll bonding but in the cases where this is impractical then adhesives of low neutron scattering cross section may be used.

2.4.4 Image Recorders

2.4.4.1 As the choice of an image recorder will depend upon the need to obtain either radiographic quality or speed, it is only possible to give general guidance as to their selection. When high quality is required a fine grain film or track-etch material should be used, when speed is the important parameter then fast X-radiographic type films should be used.

2.4.4.2 The image recorders given in the following table are recommended, based upon the practical experience of radiographers.

TABLE 2.5 IMAGE RECORDERS

Technique	Type of Recorder	Converter foils	Image recorder characteristics
Direct	Kodak-SR	100 μm Gd	Fine grain
	AGFA-D7	100 μm Gd	
Indirect	Kodak-M	100 μm Dy	Medium speed
	Kodak-SR	100 μm Dy	
(Epi-thermal neutrons)	Kodak-SR	100 μm Au	
Track-etch	CA 8015-B	Coated Lithium Borate	Low contrast
	CA 8015	Enriched ^{10}B + BN1	
	CN 85	BN1	

2.4.5 Cassettes

2.4.5.1 The cassette should be chosen to avoid backscatter and to obtain the maximum contact between the film and the converter foil, as loss of contact gives rise to image unsharpness.

2.4.5.2 Flat, rigid cassettes of the vacuum type should be used wherever possible, alternatively the compression type may be employed. Flexible cassettes should only be used when it is not possible to use the types recommended above.

2.4.5.3 The contact between the foil and the film should be tested periodically by the 'wire-mesh' method described in Appendix B of B.S. 4304: 1968 (Specification for X-Ray Film Cassettes).

2.4.6 Masking and Backscatter Protection

2.4.6.1 A significant fraction of the thermal cross section of nuclear fuels is due to scattering and thus the masking of the region surrounding the object by a neutron absorbing material can be helpful in reducing scattered radiation.

2.4.6.2 Similarly, the use of neutron absorbing materials covering the shield walls that surround the object is also recommended as this will reduce the backscattered radiation.

2.4.6.3 Backscatter can also be minimised by confining the neutron beam to the smallest practical field and by placing absorbing material behind the recording film.

2.4.6.4 If there is any doubt about the adequacy of the protection from backscattered radiation then a technique employed by X-radiography may be employed. Attach a characteristic symbol (typically a letter B) of an absorbing material to the back of the cassette and take a radiograph in the normal manner. If the image of the symbol appears on the radiograph it is an indication that the protection against backscattered radiation is insufficient.

2.4.7 Geometry

2.4.7.1 The manner in which:
a) the size of the collimator inlet aperture
b) the distance between the inlet aperture and the object, and
c) the distance between the object and the image converter
control the geometric unsharpness is fully described in Part 1 of this Handbook and it is sufficient to say here that dimensions (a) and (c) should be as small as possible and distance (b) as large as possible in order to achieve the best resolution.

2.4.7.2 Furthermore, the reciprocal relationship between these distances should be noted, in that the same fractional change in both dimensions will leave the geometric unsharpness unchanged.

2.4.7.3 It must also be recognised that the effective collimator inlet aperture size is often not the true source size due to the finite nature of the neutron source. It is therefore recommended that the true apperture size be measured by the method of measuring the collimator ratio as described by Newacheck and Underhill [Ref. 55].

2.4.8 Density of the Radiograph

2.4.8.1 In principle the amount of information that can be recorded on a radiographic film will increase with film density, and the recovery of this information will be dependant upon the ability of the viewing equipment to illuminate the image. The practical limit to this statement is a density of about 4 and in special cases such densities may be used.

2.4.8.2 However for normal radiography a density between 2 and 3 is recommended. These values are inclusive of fog and base densities of not greater than 0,3.

2.4.9 Contrast

The contrast of the film and hence its ability to discriminate a discontinuity, depends upon the:
a) variation in specimen thickness,
b) neutron energy of the beam,
c) quality of the beam e.g. the variation of neutron energies and the amount of gamma rays for the direct technique,
d) scattered radiation,
e) type of film,
f) film development and
g) film density

and their relationship are described in Part 1 of this Handbook.

2.4.10 Image Quality Indicators

2.4.10.1 An image quality indicator is a device employed to provide evidence on a radiograph that the technique that was used was satisfactory and so the use of image quality indicators given Part 3 of this Handbook is therefore recommended.

2.4.10.2 The acceptable sensitivity of the radiograph should be agreed between the purchaser and supplier based upon a recommended guide value of 2%.

2.4.11 Exposure Chart/Technique Log

2.4.11.1 It is recommended that operators of neutron radiographic facilities construct an exposure chart/technique log for the neutron radiography of nuclear fuel.

2.4.11.2 This should record the following:
a) diameter of beam inlet aperture,
b) inlet aperture – object distance (L/D ratio),
c) characteristic neutron energy (Cd ratio),
d) beam quality data as measured by a beam quality indicator,
e) description or sketch of the object set-up,
f) material(s) of the object,
g) geometry and thickness of the material(s),
h) material of the converter foil,
i) type of film,
j) film density on the image of the quality indicator,
k) identification number of radiograph,
l) exposure time,
m) details of any filter used,
n) type of developer used,
o) processing time and temperature,
p) type of image quality indicator,
q) sensitivity value measured by the image quality indicator.

2.4.12 Track-Etch Techniques

2.4.12.1 The selection and use of track etch materials is described in Part 1 of this Handbook. The recommended etching conditions for Kodak CA-8015 B, CA-8015 and CN 85 nitrocelullose film is:
- etchant, 150 g/l potasium hydroxide (KOH)
- temperature, 40°C
- time, 30 min.

2.4.12.2 It is recommended that, in order to achieve a strict temperature control of the bath it should be heated in a furnace and stirred before use. Long etching times should be avoided in order to avoid sediment formation in the bath due to the camfer removed from the nitrocelullose. Agitation during the etching period causes cloudiness on the nitrocelullose film and should therefore be avoided.

2.4.12.3 When track etch materials are being used then items (h) and (i) in the list at 2.4.11.2 will be modified as follows:
h1) type of track etch converter
h2) type of track etch material
i) etching time/temp.

2.5 MEASUREMENT

2.5.1 Definition and Methods

2.5.1.1 In the context of this document measurement may be defined as the determination of the physical size of some feature of a fuel pin or similar object, i.e. fuel pellet diameter or length, radial gaps, cladding thickness, etc.

2.5.1.2 Measurement may be made directly from the radiograph, making due allowance for any enlargement or reduction caused by the radiographic conditions, or by the use of a comparitor of known dimensions which also appears on the radiograph.

2.5.1.3 As this document is only concerned with the radiography of nuclear fuel the following discussion will be confined to the measurement of cylindrical object.

2.5.2 The Principles of Radiographic Measurement

The principles of radiographic measurement are described in Part 1 of this Handbook and it is sufficient to say here that the accuracy of a radiographic measurement technique is dependant upon the sharpness of the image and the contrast. The following recommendations therefore aim at optimising the sharpness and the related contrast of the image and proposes various methods of enhancing the image and taking dimensional measurement from it.

2.5.3 The Neutron Radiographic Technique

As the object of radiographic measurement of nuclear fuel is to make a quantitative evaluation of the results of irradiation then the object will be radioactive and hence a transfer technique must be used. The following discussion will therefore assume the use of the transfer technique, whilst accepting that for non-irradiated specimens it may be convenient to make some exposures by the direct technique.

2.5.4 Making the Radiograph

2.5.4.1 Every precaution should be taken to ensure a sharp image of adequate contrast, by:

a) elimination of all relative movement of the object and the image converter recorder combination,
b) using a high geometric sharpness,
c) using a high resolution image recorder,
d) using a high resolution converter foil,
e) optimising the neutron energy and image converter relationship,

f) ensuring that the beam is well collimated,
g) using a vacuum cassette,
h) avoiding back scatter,
i) careful preservation and handling of the image recorder and films,
j) avoidance of fogging on photographic image recorders,
k) careful development techniques.

2.5.4.2 When the radiograph has been produced it should be kept in a protective envelope at all times and under storage conditions recommended by the manufacturer.

2.5.5 Making the Measurements

2.5.5.1 The following sections give, where possible, data in support of the items listed in 2.5.4.1 above. This data has been extracted from the references given in Part 1 of this Handbook. The following is therefore a summary of the practices used by experienced radiographers and is not necessarily well supported by a complete theoretical understanding. It may also be dependant upon the characteristics of the neutron radiography equipment in use.

2.5.5.2 In making these recommendations it is recognised that the final result is dependant upon the combined effect of all the above variables, and so it is of little use to devote resources, say, to achiving a very high geometric resolution when the resolution of the image recorder is very poor. The problem of determining how much improvement should be made to any particular aspect of the radiographic system can only be resolved by measuring the transfer function of each component in the system, and as this is difficult and costly, it is normally beyond the scope of practicing radiographers.

2.5.5.3 The data given below should therefore be used with the above reservation in mind as it does not represent an optimum set of conditions, but only a consensus of opinion.

2.5.5.4 **Vibration** can be a problem when there are machines (e.g. cranes etc.) is use in nearby buildings. This should be verified by taking both short and long exposures of the object with a camera, using a slow speed photographic film, with the camera mounted on a base that is relatively unaffected by the vibrations.

2.5.5.5 **Geometry.** The collimator ratio (L/D) should be 100 or higher, but it is considered that the advantages of increasing the ratio greater than 300 are diminishing.

2.5.5.6 **Converter foils** for the transfer method are limited to indium dysprosium, and gold, all of which emit a particle of approximately 1 MeV, i.e. long range and not conducive to high resolution. However, the dysprosium foils are thinner and therefore have better resolution capability. A thickness of 0,025 mm or less is recommended.

2.5.5.7 **Image recorders** to be used for measurement are film or celulose acetate. Films are discussed in para. 2.4. Celulose acetate has the higher resolution, but very low contrast. It is recommended that an increase in contrast is obtained by copying the original on to Kodalith film type 2571 by means of a point source, or condenser type, photographic enlarger.

2.5.5.8 **Neutron energy and image converter combination.** It is recommended that indium, and dysprosium converters are used with thermal neutrons and indium and gold converters for epithermal neutrons.

2.5.5.9 **Collimation** is dependant upon the L/D ratio and this is discussed in para. 2.5.5.5. It is also dependant upon the detail design of the collimator and this is described in Part 1 of this Handbook. It is recommended that a beam quality indicator should be used to measure the characteristics of the beam and the values given in part 3 of the handbook are recommended.

2.5.5.10 **Cassettes** of the vacuum type are recommended.

2.5.5.11 **Backscatter** should be measured by the method given in para. 2.4.6.4.

2.5.5.12 **Preservation and handling of the converter foils and films** should follow an established routine using the following recommendations:
a) store in a container that preserves the surface condition and the flatness,
b) never handle the image recording surface,
c) ensure that the previous image is fully decayed before re-use,
d) keep the recording surfaces clean and bright,
e) the recommendations of para. 2.7.3 on handling should be followed.

2.5.5.13 **Fogging** of photographic films may be avoided by checking that, cassettes are fully light-tight and that the recommendations of Section 2.7 are followed.

2.5.5.14 **Development techniques** given in Section 2.8 should be followed.

2.5.6 Image Enhancement

2.5.6.1 Electronic Methods

Some advantages can be gained by using electronic enhancement systems to improve the contrast and resolution at the edge of a specimen or internal feature. An iterative process is usually required. However, care must be taken to ensure that the results so obtained are meaningful by making frequent reference to image quality indicators or the dimensions of reference features within the radiograph.

2.5.6.2 **Optical Methods**

Improvements can be made by magnifing the image by optical projection. A magnification of up to 10x is recommended.

2.6 SAFETY PRECAUTIONS

2.6.1 Whenever a neutron radiography facility is in use it is essential that adequate precautions are taken to protect the operator and other persons in the vicinity from uncontrolled exposure to radiation.

2.6.2 It is recommended that these precautions should adhere to the local safety rules and that there should be a written procedure describing every type of neutron radiographic technique in use and the individual steps in each technique. This procedure should include the health physics controls that shall be applied, as agreed with the local area Health Physics Officer.

2.6.3 The responsibility for following the procedure shall be clearly stated in writing and it is recommended that the person responsible for Health Physics Control shall make regular audits to ensure that the procedure is being followed.

2.7 FILM HANDLING

2.7.1 Storage of Film

Unexposed films should be stored in such a manner that they are protected from the effects of light, pressure, excessive heat, excessive humidity, damaging fumes or vapours, or penetrating radiation. Film manufactures should be consulted for detailed recommendations on film storage. Storage of film should be on a 'first in', 'first out' basis.

2.7.2 Safelight Test

Films should be handled under safelight conditions in accordance with the film manufacturer's recommendations.

2.7.3 Cleanliness and Film Handling

2.7.3.1 Cleanliness is one of the most important requirements for good radiography. Cassettes and screens must be kept clean, not only because dirt retained may cause exposure or processing artifacts in the radiographs, but because such dirt may also be transferred to the loading bench and subsequently to other films or screens.

2.7.3.2 The surface of the loading bench must also be kept clean.

2.8 FILM PROCESSING

2.8.1 General

To produce a satisfactory radiograph, the care used in making the exposure must be followed by equal care in processing. The most careful radiographic techniques can be nullified by incorrect or improper darkroom procedures.

2.8.2 Automatic Processing

The essence of the automatic processing system is control. The processor maintains the chemical solutions at the proper temperature, agitates and replenishes the solutions automatically and transports the films mechanically at a carefully controlled speed troughout the processing cycle. Film characteristics must be compatible with processing conditions. It is, therefore, essential that the recommendations of the film, processor and chemical manufacturers be followed.

2.8.3 Manual Processing

2.8.3.1 This section outlines the steps for one acceptable method of manual processing. Modifications, provided they are shown to be adequate, may also be used.

2.8.3.2 **Preparation** – No more film should be processed than can be accomodated with a minimum separation of 12 mm. Hangers are loaded and solutions stirred before starting development.

2.8.3.3 **Start of Development** – Start the timer and place the films into the developer tank. Separate to a minimum distance of 12 mm and agitate in two directions for about 15 s.

2.8.3.4 **Development** – Normal development is 5 to 8 min at 20° C. Longer development time generally yields faster film speed and slightly more contrast. The manufacturer's recommendations should be followed in choosing a development time. When the temperature is higher or lower, development time must be changed. Again, consult manufacturer-recommended development time versus temperature charts. Other recommendations of the manufacturer to be followed are replenishment rates, renewal of solutions and other specific instructions.

2.8.3.5 **Agitation** – Shake the film horizontally and vertically, ideally for a few seconds each minute during development. This will help film develop evenly.

2.8.3.6 **Stop Bath or Rinse** – After development is complete, the activity of developer remaining in the emulsion should be neutralised by an acid stop bath or, if this is not possible, by rinsing with vigorous agitation in clear water. Follow the film manufacturer's recommendation of stop bath composition (or length of alternative rinse), time immersed and life of bath.

2.8.3.7 **Fixing** – The films must not touch one another in the fixer. Agitate the hangers vertically for about 10 s and again at the end of the first minute, to ensure uniform and rapid fixation. Keep them in the fixer until fixation is complete (that is, at least twice the clearing time), but not more than 15 min in relatively fresh fixer. Frequent agitation will shorten the time of fixation.

2.8.3.8 **Fixer Neutralising** – The use of a hypo eliminator or fixer neutraliser between fixation and washing may be advantageous. These materials permit a reduction of both time and amount of water necessary for adequate washing. The recommentations of the manufacturers as to preparation, use and useful life of the baths should be observed rigorously.

2.8.3.9 **Washing** – The washing efficiency is a function of wash water, its temperature and flow and the film being washed. Generally washing is very slow below 16° C. When washing at temperatures above 30° C, care should be excercised not to leave films in the water too long. The films should be washed in batches without contamination from new film brought over from the fixer. If pressed for capacity, as more films are put in the wash, partially washed film should be moved in the direction of the inlet.

2.8.3.10 The cascade method of washing uses less water and gives better washing for the same length of time. Divide the wash tank into two sections (maybe two tanks). Put the films from the fixer in the outlet section to the inlet section. This completes the wash in the fresh water.

2.8.3.11 For specific washing recommendations, consult the film manufacturer.

2.8.3.12 **Wetting Agent** – Dip the film for approximately 30 s in a wetting agent. This makes water drain evenly off film which facilitates quick, even drying.

2.8.3.13 **Fixer Concentrations** – If the fixing chemicals are not removed adequately from the film they will in time cause staining or fading of the developed image. Permissible residual fixer concentrations depend upon whether the films are to be kept for commercial purposes (3 to 10 years) or must be of archival quality. Archival quality processing is desirable for all radiographs whenever average relative humidity and temperature are likely to be excessive, as is the case in tropical and subtropical climates. The method of determining residual fixer concentrations may be ascertained by reference to ANSI PH4.8., PH1.28, PH4.32 and PH1.41.

2.8.3.14 **Drying** – Drying is a function of (1) film (base and emulsion); (2) processing (hardness of emulsion after washing, use of setting agent); and (3) drying air (temperature, humidity, flow). Manual drying can vary from still air drying at ambient temperature to as high as 60° C with air circulated by a fan. Film manufacturers should again be contacted for recommended drying conditions. Take precaution to tighten film on hangers so that it cannot touch in the dryer. Too hot drying temperature at low humidity can result in uneven drying and should be avoided.

2.8.3.15 It is desirable to monitor the activity of the radiographic developing solution. This can be done by periodic development of film strips exposed under carefully controlled conditions, to a graded series of radiation intensities or time, or by using a commercially available strip carefully controlled for film speed and latent image fading.

2.9 VIEWING RADIOGRAPHS

2.9.1 The illuminator must provide light of an intensity that will illuminate the average density areas of the radiographs without glare and it must diffuse the light evenly over the viewing area. Commercial fluorescent illuminators are satisfactory for radiographs of moderate density; however, highintensity illuminators are available for densities up to 3,5 or 4,0. Masks should be available to exclude any extraneous light from the eyes of the viewer when viewing radiographs smaller than the viewing port or to cover low-density areas. Viewing radiographs requires considerable handling; therefore, it is recommended that films be handled with extreme caution.

2.9.2 Subdued lighting, rather than total darkness, is preferable in the viewing room. The brightness of the surroundings should be about the same as the area of interest in the radiograph. Room illumination must be so arranged that there are no reflections from the surfaces of the film under examination.

2.10 REFERENCE RADIOGRAPHS

Part 4 of this Handbook consists of a collection of reference radiographs which show defects in nuclear fuel. It is recommended that these radiographs be used when making interpretations and that whenever possible the applicable reference radiograph number should be quoted in the report on the interpretation.

2.11 STORAGE OF RADIOGRAPHS

Radiographs should be stored using the same care as for any other valuable record. Envelopes having an edge seam, rather than a centre seam and joined with a nonhygroscopic adhesive, are preferred, since occasional staining and fading of the image is caused by certain adhesives used in the manufacture of envelopes (see ANSI PH4.20).

2.12 RECORDS AND REPORTS

2.12.1 Records

It is recommended that a work log (a log may consist of a card file, punched card system, a book, or other record) constituting a record of each job performed, be maintained. This record should comprise, initially, a job number (which should appear also on the films), the identification of the parts, material or area radiographed, the data the films are exposed and a complete record of the radiographic procedure, in sufficient detail so that any radiographic techniques may be duplicated readily. If calibration data, or other records such as card files or procedures, are used to determine the procedure, the log need refer only to the appropriate data or other record. Subsequently, the interpreter's findings and disposition (acceptance or rejection), if any, and his intials, should also be entered for each job.

2.12.2 Reports

When written reports or radiographic examinations are required they should include the following, plus such other items as may be agreed upon:

a) Identification of parts, material or area.
b) The radiographic job number.
c) The findings and disposition, if any. This information can be obtained directly from the log.

3. NRWG INDICATORS FOR TESTING OF BEAM PURITY, SENSITIVITY, AND ACCURACY OF DIMENSIONS OF NEUTRON RADIOGRAPHS

For the sake of testing the radiographic image quality and accuracy of dimension measurements from neutron radiographs of reactor fuel, the NRWG has decided to produce and test special indicators developed for that purpose.

In the preliminary investigation it was determined that there are no suitable indicators prescribed in the existing standards on neutron radiography. The only published standard in that field [Ref. 1], the ASTM E545-75, was prepared for general neutron radiography and is now under revision.

Taking into account the work done on this revision (as e.g. described in [Ref. 2]) as well as different proposals made by the NRWG members [Refs. 3, 4, 5], it was decided to produce the following indicators for neutron radiography of nuclear fuel:

- Beam Purity Indicator (BPI)
- Beam Purity Indicator-Fuel (BPI-F)
- Sensitivity Indicator (SI)
- Calibration Fuel Pin (CFP-E1)

Those indicators, fabricated at Risø National Laboratory *), were distributed among all NRWG participants and will be tested under a special NRWG Test Program [Ref. 6]. The design of the above-mentioned indicators is described below.

It is worth noting that some work is going on in the NRWG on the development of a common Sensitivity and Measurement Indicator-Fuel (SMI-F) and a Combined Quality Indicator (QIF), as described in [Ref. 4]. Those indicators are not yet included within the present Test Program [Ref. 6].

3.1 THE VARIOUS INDICATORS

3.1.1 Beam Purity Indicator (BPI)

The neutron beam and image system parameters that contribute to film exposure and thereby affect overall image quality can be assessed by the use of Beam Purity Indicators. Following the experience gained during the use of the BPI prescribed by the first ASTM standard on neutron radiography [Ref. 1] a new BPI design was developed, which will be recommended by the revised ASTM standard.
This design , shown on Fig. 3.1, was adopted by the NRWG, and will be tested under its Test Program [Ref. 6].

*) *on behalf of the Petten Establishment of the Joint Research Centre of the Commission of the European Communities.*

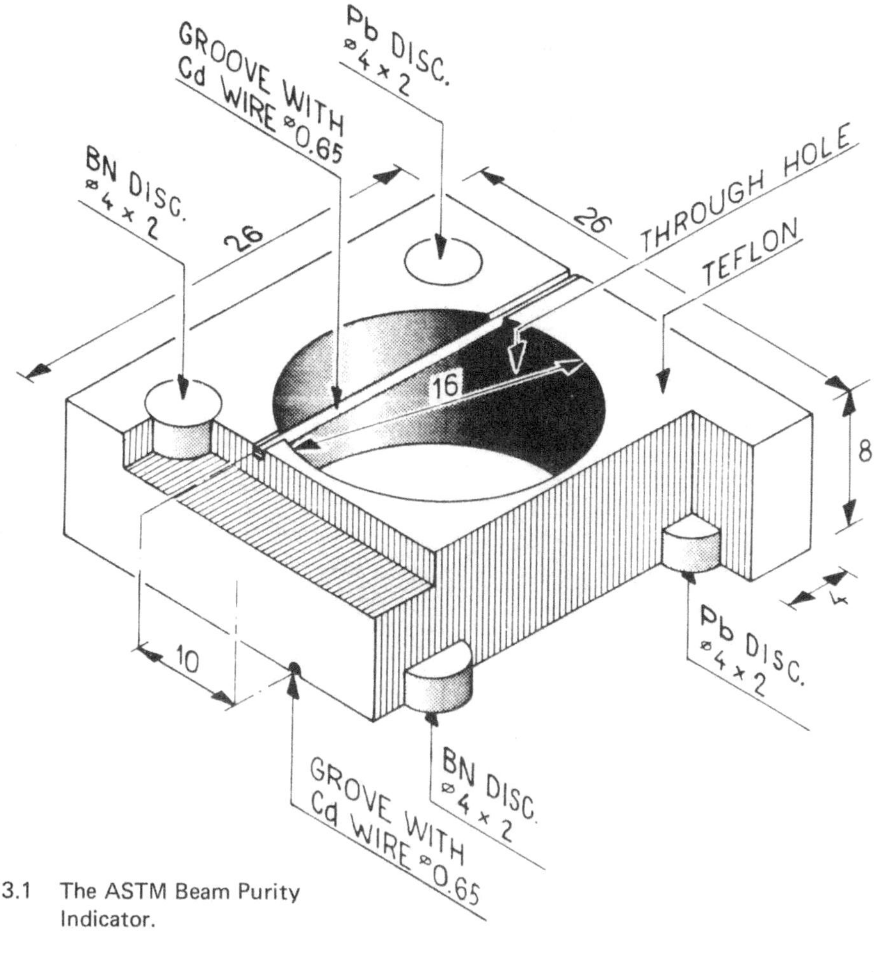

Fig. 3.1 The ASTM Beam Purity Indicator.

The body of the BPI is made of a 8 mm thick teflon (26 x 26 mm) plate. It has a central hole of 16 mm in diameter. In the teflon plate two grooves to accommodate 0,64 mm cadmium wires are made, separated by 10 mm from each other. At the top and bottom of the teflon plate two holes, 4 mm in diameter and 2 mm deep, are machined. At each side of the BPI a boron nitride and a lead disc (2 mm thick) are inserted into the circular holes.

Key feature of the device is the ability to make a visual analysis of its image for subjective quality information. Densitometric measurements of the image of the device permit quantitative determination of radiographic contrast, low energy gamma contribution, pair production contribution, image unsharpness, and information regarding film and processing quality.

To be able to identify the orientation of the BPI on neutron radiographs, one corner of the indicator was cut off (not shown on Fig. 3.1).

3.1.2 Beam Purity Indicator-Fuel (BPI-F)

For controlling the neutron beam components in nuclear fuel radiography the NRWG has developed a special Beam Purity Indicator-Fuel, which is a modification of the ASTM BPI (See. Fig. 3.2).

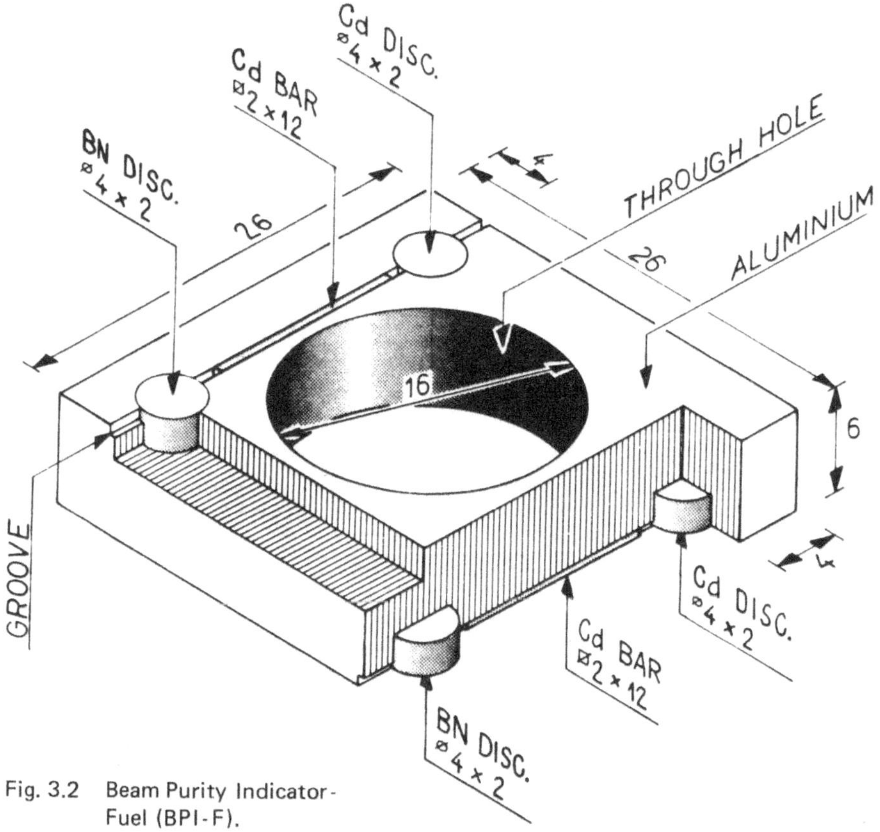

Fig. 3.2 Beam Purity Indicator-Fuel (BPI-F).

The body of the BPI-F consists of a 6 mm thick aluminium plate (26 x 26 mm^2), in which a 16 mm round central hole is machined. At the top and bottom of the Al plate two pairs of round holes (4 mm in diameter and 2 mm deep) are made to accommodate 2 mm thick boron nitride and cadmium discs. Through those holes square grooves (2 x 2 mm^2) are machined to accommodate 12 mm long square (2 x 2 mm^2) cadmium bars.

The reasons behind the modification of the ASTM BPI are explained in [Ref. 3] as follows :
"The materials of the ASTM BPI were principally chosen to be suitable for the detection of gamma rays and as it is assumed that when the BPI-F is in use, a transfer or track etch technique will be used, clearly a sensitivity to gammas is not needed. It is therefore considered that the base material should be aluminium and that the filter-discs should be boron nitride and cadmium (the ASTM design has boron nitride and lead discs)".

To be able to identify the orientation of the BPI-F on neutron radiographs one corner of the indicator was cut off (not shown on Fig. 3.2).

From measurements of film densities under different parts of the BPI-F, and background density, different neutron beam components can be calculated.
The cadmium wires or rods included in each beam purity indicator are used to provide an indication of inherent beam resolution or sharpness.

3.1.3 Sensitivity Indicator (SI)

Instead of the former four types of ASTM Sensitivity Indicators [Ref. 1] one new type of SI was developed (Fig. 3.3).
This sensitivity indicator basically combines a hole gauge and gap gauge into a small single device. The holes are sized to be smaller than can be seen by conventional neutron radiography, and they progress up in size. Similarly, the gaps formed by aluminium shims between sheets of acrylic resin cover a range that is useful for all facilities.

The NRWG has considered a special design of a sensitivity indicator, including steps and shims of UO_2, which could be useful in evaluating the image quality of neutron radiographs of nuclear fuel. Unfortunately, it is technically not feasible to construct such an indicator and therefore the ASTM SI was adopted by the NRWG for its Test Program.

3.1.4 Calibration Fuel Pin (CFP-E1)

As mentioned in [Ref. 2] ; "The design goal for the ASTM sensitivity indicator is to provide the maximum sensitivity information in an easy to manufacture and easy to interpret configuration. It is recognized that the only true valid sensitivity indicator is material or component, equivalent to the part being neutron radiographed, with a known standard discontinuity (reference standard comparison part)".

Such a "reference standard comparison part" for nuclear fuel pins is the calibration fuel pin CFP-E1(Fig. 3.4). It is described in [Ref. 7]. According to the specifications given in [Ref. 7] ten calibration fuel pins were produced at Risø and distributed among the NRWG members to be tested under the Test Program [Ref. 6].

The calibration fuel pin CFP-E1(Fig. 3.4) incorporates the following features :

- From the nine UO_2 pellets two are made of natural, and seven of enriched uranium.
- All the pellets have a different length.
- The two pellets made of natural uranium and one pellet of enriched uranium have a constant diameter on all their lengths, to fit closely into the zircaloy cladding tube (practically no fuel-to-cladding gaps).
- The remaining six UO_2 pellets of enriched uranium have a reduced diameter on half of their lengths so as to form a calibrated fuel-to-cladding gap. These radial gaps are 50, 100, 150, 200, 250 and 300 μm wide.
- The first UO_2 pellet from natural uranium and the first pellet of enriched uranium have a dishing 0.3 mm deep on the surfaces facing each other.
- There are aluminium spacers between all UO_2 pellets from enriched uranium. They are simulating the pellet-to-pellet gaps. The thicknesses of those spacers are the same as the fuel-to-clad gaps, i.e. 50, 100, 150, 200, 250 and 300 μm respectively.

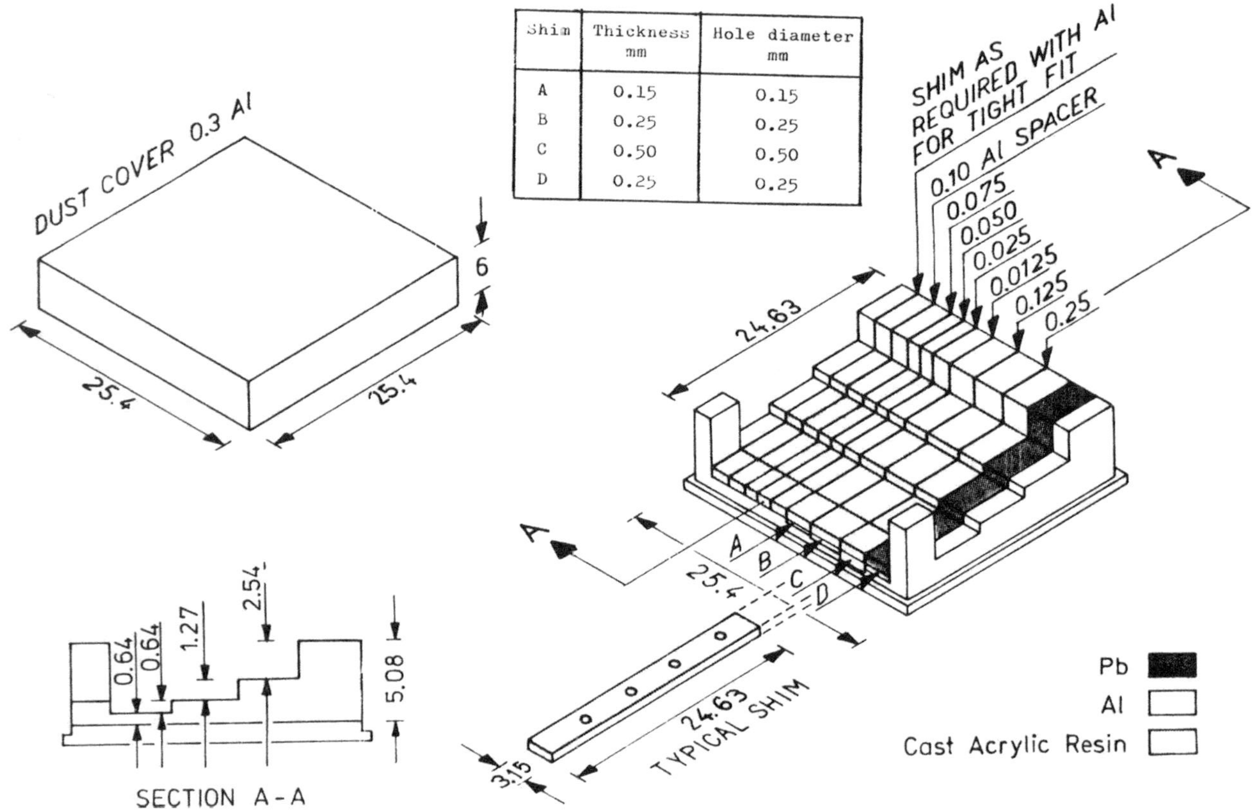

Shim and hole dimensions

Shim	Thickness mm	Hole diameter mm
A	0.15	0.15
B	0.25	0.25
C	0.50	0.50
D	0.25	0.25

Fig. 3.3 The ASTM Sensitivity Indicator.

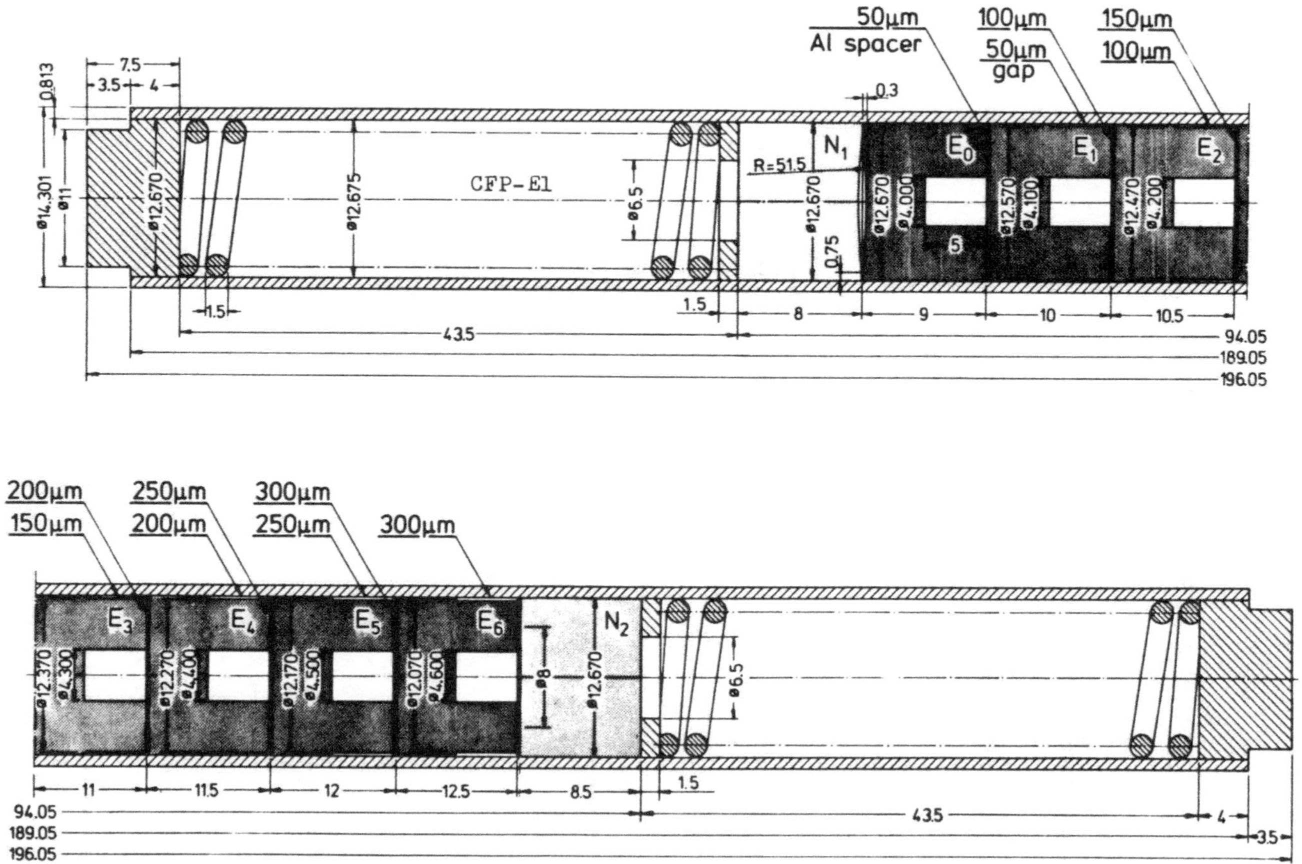

Fig. 3.4 Calibration Fuel Pin (CFP-E1).

- All UO_2 pellets made of enriched uranium have a calibrated central void. The diameter of this void is 4 000 μm increasing by an increment of 100 μm throughout the consecutive pellets to a diameter of 4 600 μm, respectively.

3.2 ASSESSMENT OF TEST RESULTS FOR THE INDICATORS

3.2.1 Assessment for the Beam Purity Indicator (BPI)

From the neutron radiographs of the BPI, the following film densities are to be measured:

- D_1 - density under the lower boron nitride disc
- D_2 - density under the upper boron nitride disc
- D_3 - density under the lower lead disc
- D_4 - density under the upper lead disc
- D_5 - background film density in the center of the hole
- D_6 - film density through the teflon body.

From those values the neutron exposure contributions can be calculated as follows :

Thermal neutron content - C

$$C = \frac{D_5 - (\text{higher value of } D_1 \text{ and } D_2)}{D_5} \times 100$$

Scattered neutron content - S

$$S = \frac{D_1 - D_2}{D_5} \times 100$$

Gamma content - γ

$$\gamma = \frac{D_6 - (\text{lower value of } D_3 \text{ and } D_4)}{D_5} \times 100$$

Pair production contribution - P

$$P = \frac{D_3 - D_4}{D_5} \times 100$$

The film density shall be measured using a diffuse transmission densitometer. The densitometer shall be accurate to ± 0.04 and repeatable to ± 0.02 density units.

Besides the above-mentioned density measurements and calculations from the radiograph of the BPI one shall further visually compare the images of the cadmium rods in the beam purity indicator. An obvious difference in image sharpness indicates an L/D ratio which is probably too low for general inspection. Detailed analysis of the rod images is possible using a scanning microdensitometer.

3.2.2 Assessment for the Beam Purity Indicator-Fuel (BPI-F)

From the neutron radiographs of the BPI-F, the following film densities are to be measured :

- D_D - density under the lower boron nitride disc
- D_B - background film density in the center of the hole
- D_C - density under the upper boron nitride disc
- D_E - density under the upper cadmium disc
- D_F - density under the lower cadmium disc.

From those values, exposure contributors can be calculated as follows :

Thermal neutron content - TN

$$TN = \frac{D_B - D_E}{D_B} \times 100$$

Epithermal neutron content -EN

$$EN = \frac{D_E - D_C}{D_B} \times 100$$

Scattered neutron content - SN

$$SN = \frac{D_C - D_D}{D_B} \times 100$$

Besides the above mentioned density measurements and calculations from the radiographs of the BPI-F, inherent and total unsharpness can be determined.

3.2.3 Assessment for the Sensitivity Indicator (SI)

The purpose of the sensitivity indicator is to determine the sensitivity of details visible on the neutron radiograph by evaluating the neutron radiographic image of the SI.

Besides one shall visually inspect the image of the lead steps in the sensitivity indicator. If the 0,25 mm holes are not visible, the exposure contribution from gamma radiation is very high and further analysis should be made.

The lead steps are shown on Fig. 3.3; under the steps a 0,25 mm thick acrylic shim D is located with four 0,25 mm holes.
When examining the neutron radiographs of the SI one shall visually inspect the image of the cast acrylic resin steps and note all the holes visible to the observer (consecutive holes marked as H). Then one shall take as the value of H reported the largest consecutive value of H that is visible in the image.

The cast acrylic resin steps, shown on the left side of the SI (see Fig. 3.3) are separated by aluminium spacers with thickness (gap size) marked as G.

During the visual examination of the neutron radiograph of the SI one shall report the G value. The value of G reported is the smallest gap which can be seen at all absorber thicknesses.

3.2.4 Assessment for the Calibration Fuel Pin (CFP-E1)

From the neutron radiographs of the CFP-E1 the following dimensions ought to be determined (see Fig. 3.4) :

Axial dimensions (read along the longitudinal axis of the pin)

- Total fuel stack length (from the beginning of pellet N_1 to the end of pellet N_2).
- Length of all pellets separately.
- Length of the central void.
- Dishing between pellets N_1 and E_0.
- Pellet-to-pellet gaps.

Radial dimensions

- Pellet diameters of nonstepped pellets (measured in the middle of the pellets N_1, E_0 and N_2).
- Pellet diameters of stepped pellets (measured in the middle of the non-stepped and in the middle of the stepped half of each pellet).
- Pellet-to-pellet gaps (both gaps at each pellet).
- Cladding tube wall thickness (measured at the same radius as the diameter and gap measurements).
- Central void diameter (measured in the middle of the void length).

All the above-mentioned measurements shall be performed using those measuring instruments (e.g. scanning microdensitometer, projection microscope) available at the various centers.

As described above, from neutron radiographs of the CFP-EI both axial as well as radial dimensions can be read. The results of those measurements shall be compared with the true dimensions as given in the CFP-E1 certificate.

3.3 REFERENCES

1. ASTM E545-75. Standard method for determining image quality in thermal neutron radiographic testing.

2. R.L. Newacheck :
The ASTM neutron radiography IQI system. Theory of operation and accuracy. Aerotest Operations Inc., February 1980.

3. R.S. Matfield :
On the problem relating to the accuracy of the measurement of fuel pin diameters by neutron radiography.
ES. 4.682.A.012, 03.07.1980.

4. R.S. Matfield :
On the design of image quality indicators for the neutron radiography of nuclear fuel pins.
ES. 4.682.A.014, 16.02.1981.

5. J.C. Domanus :
Accuracy of dimension measurements from neutron radiographs of nuclear fuel pins.
Risø-M-1860, 26.03.1976 and Eighth World Conference on Nondestructive Testing, Cannes 6 - 11.09.1976, Paper 3L-8.

6. J.C. Domanus :
Revised test program for testing of the CFP-EI; ASTM (revised) BPI and SI and BPI-F.
Risø Report B-512, August 1981.

7. J.C. Domanus :
Calibration fuel pin CFP-EI. Risø National Laboratory, Metallurgy Department. Report B-511, August 1981.

4. ATLAS (COMPACT VERSION) OF DEFECTS REVEALED BY NEUTRON RADIOGRAPHY IN LIGHT WATER REACTOR FUEL

4.1 INTRODUCTION

The assessment of neutron radiographs of nuclear fuel pins can be done much easier, faster and simpler if reference can be made to typical defects, which can be revealed by neutron radiography. In the fields of industrial γ-radiography such collections of reference radiographs, showing typical defects in welding, or casting have been compiled and published some time ago.

Since the early 1970's neutron radiography is routinely used for the quality and performance control of nuclear fuel. During the assessment of neutron radiographs, some typical defects of the fuel were found and it was felt that a classification of such defects would help to speed up the assessment procedure. Therefore, in the frame of the programme of the Neutron Radiography Working Group, an atlas of reference neutron radiographs has been compiled [Ref. 1], which was printed as a working document on behalf of JRC Petten in June 1979. It contains a collection of typical defects revealed by neutron radiography in light water reactor fuel, which are reproduced on X-ray film (original size) and as enlargements (2x) on photographic paper.

A revised version of the atlas, which is supplemented with further examples of typical defects is under preparation and will be edited by the Neutron Radiography Working Group.
It was not possible to reproduce in the handbook all the neutron radiographs contained in the atlas. Therefore a selection was made of those enlargements which illustrate the most characteristic defects occurring in light water reactor fuel.

4.2 RELEVANT NOTES

4.2.1 Fuel Pins

For the purpose of the present collection of neutron radiographs a typical example of a nuclear fuel pin , used in light water reactors, was chosen.
Fig. 4.1 shows all the components of such a fuel pin where defects, detectable by neutron radiography, can occur.

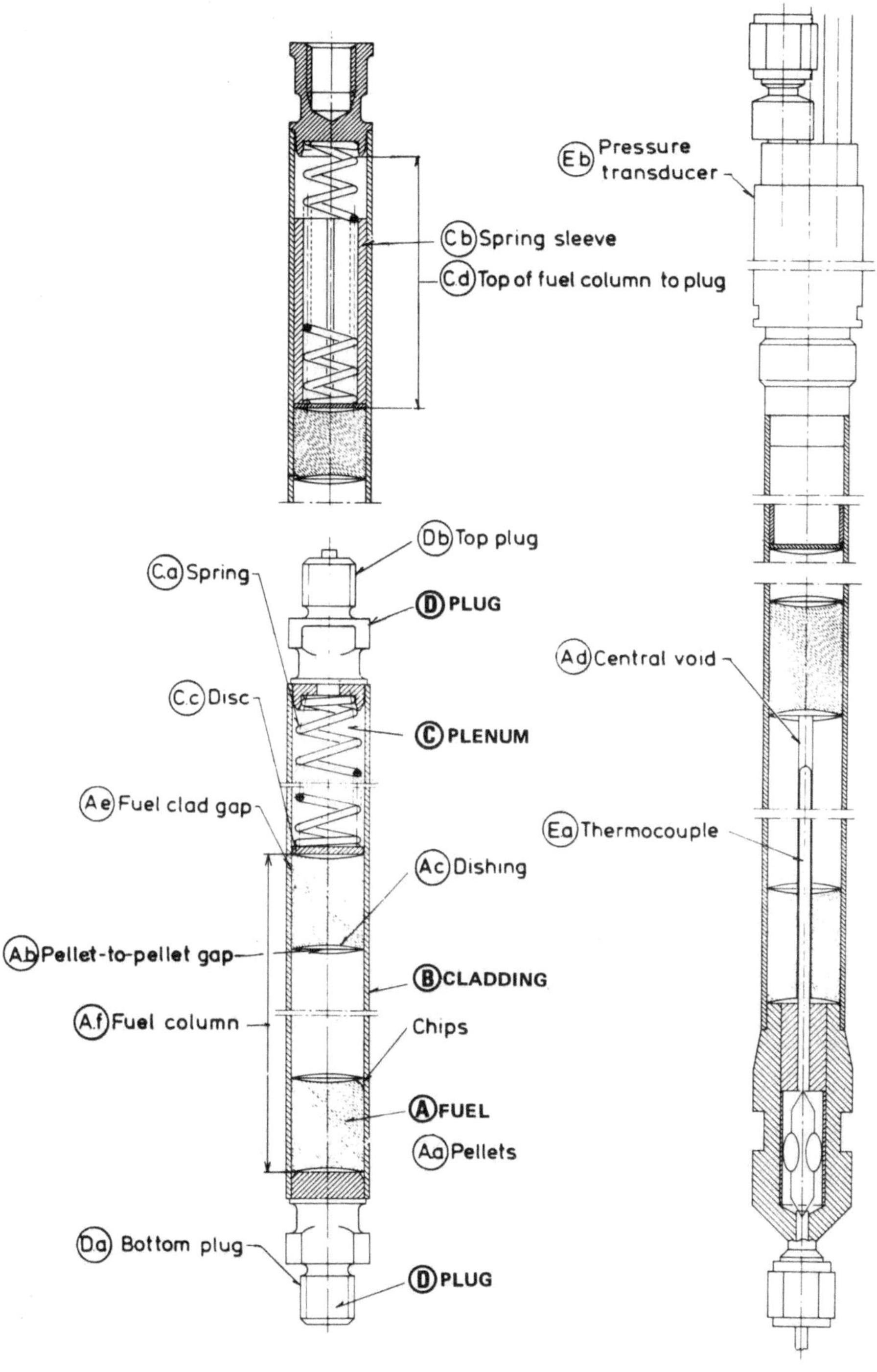

Fig. 4.1 Components of a typical nuclear fuel pin.

Those components are marked with capital letters as follows:

- Nuclear fuel : "A"
- Fuel Cladding : "B"
- Plenum : "C"
- End plugs : "D"
- Instrumentation : "E".

4.2.2 Defect

In the present collection of neutron radiographs the term "defect" is used for designation of a neutron radiographic finding, showing a different appearance of a particular part of the fuel, different from that, which will be shown on a neutron radiograph of that part as fabricated. The term "defect" is therefore used in a rather general and neutral significance. A "defect" in the sense of this Handbook does not necessarily disqualify a fuel pin for further normal operation.

4.2.3 Defect Location

On Fig. 4.2 the fuel pin components shown on Fig. 4.1 are subdivided into elements where defects may occur (listed in the vertical column at the left and marked with small letters).

LEGEND:

- ● Defects shown on display.
- ○ Defects occuring, but not shown.
- × Dimensions from neutron radiographs.
- ⊗ Dimensions measured at Risø.

Defect intensity: 1 - small, 2 - medium, 3 - high

DEFECT LOCATION:			1	2	3	4	5	6	7	8	9	10	11	12	13	14	15	16	17	18	19	20	21	22	23	24
			Cracks				Chips																	Dim.		
			Random	Longitudinal	Transverse	Annular	Corner	Other	In pellet-to-pellet gap	In fuel-to-clad gap	Not present	Enlarged	Contracted	Filled up	Deformed	In one pellet	Through sev. pellets	Through whole fuel column	Differerent n-absorption	Hydrides	Broken	Dislocated	Expanded	Diameter	Thickness	Length
A Fuel	a	Pellets	●	●	●	●	●	●	●	○	●	●							●		●			×		×
	b	Pellet-to-pellet gap										●	●													×
	c	Dishing												●	●											
	d	Central void														●	●	●						×		×
	e	Fuel-to-clad gap										●	○												×	×
	f	Fuel column																								⊗
B	a	Cladding													●					●	●				×	
C Plenum	a	Spring											●		●						○	●	○	×	×	×
	b	Spring sleeve													○					○	●	○		×	×	×
	c	Disc													○					○	○	●		×	×	
	d	Top of fuel column to plug																								⊗
D Plug	a	Bottom end													○					●	○	○		×		×
	b	Top end													○					●	●	○		×		×
E Instrum.	a	Thermocouple													○						○	●		×		×
	b	Pressure transducer													○						○	○		×		×
	c	Other													○						○	○		×		×

Fig. 4.2 Fuel pin components and defects occuring in them.

4.2.4 Defect Nature and Origin

Defects which may occur in different elements of the fuel pins can be of different nature and origin. They are listed at the top of Fig. 4.2 (columns 1 to 21).

4.2.5 Defect Occurrence

On Fig. 4.2 the sign "•" signifies, that in that location a particular defect can occur and that this defect is illustrated in the present collection.
There are, however, more defects which can most likely occur in nuclear fuel and can be detected by neutron radiography, but which are not found among the radiographs of the Atlas. They are marked "o" on Fig. 4.2.

4.2.6 Defect Intensity

Defects in nuclear fuel can occur with different intensity (e.g. cracks in fuel pellets can be miniscule, slightly visible, or so big as to break the whole pellet). Therefore it was felt that one shall also classify the intensity of the defects. For that purpose an arbitrary three grade scale was adopted: 1 - meaning small, 2 - medium and 3 - high intensity defect. This intensity classification is used routinely for the assessment of defects revealed by neutron radiography.

4.2.7 Dimensions

It is also possible to measure dimensions from neutron radiographs. Therefore the last three columns (22 to 24) at the top of Fig. 4.2 list those dimensions.

4.2.8 Measuring of Dimensions

Besides the defects, dimensions of various elements of the fuel pins can be determined from neutron radiographs. Those instances are marked "x" on Fig. 4.2 and those which are routinely measured during the assessment of neutron radiographs are marked with "⊠".

4.3 THE COLLECTION OF THE ATLAS

4.3.1 Contents of the Collection in Ref 1.

The collection of the Atlas contains neutron radiographs of defects marked with "o" on Fig. 4.2 (see also chapter 4.2.5).

The original neutron radiographs were taken at the DR1 RISØ reactor (2 kW) on double coated Agfa Gevaert Structurix D4 X-ray film. A transfer technique was used with a 0,1 mm dysprosium foil.
Exposure time was about 30 min. to a $1{,}6 \times 10^6$ $n.cm^{2}.s^{-1}$ neutron beam ($10 \times 10 cm^2$).
The L/D ratio in the vertical direction of the neutron beam (perpendicular to the fuel pin axis) was 110 and in the horizontal direction (coinciding with the pin axis) was 27,5.

The radiographs in the Atlas are reproductions of the original neutron radiographs copied on Kodak X-Omat Duplicating Film.
The original neutron radiographs were also photographed on a 35 mm Agfapan 100 film and thereafter enlarged (2x) on photographic paper. A selection of these enlargements is also included in the present publication.

4.3.2 The Use of the Collection in Ref. 1

The copies of the neutron radiographs on film can be viewed without removing them from the Atlas, because there is a blank page following each copy. This blank page can be illuminated by a shaded desk lamp.
If necessary the reference radiograph may be removed from the collection to be viewed on an illuminator together with the radiograph under assessment for comparison.

4.3.3 The Selection of Characteristic Defects

A selection of defects revealed by neutron radiography in light water reactor fuel is given below. Enlargements (magn. 2x) of neutron radiographs on photographic paper are reproduced.
The defects' location and their nature and origin are marked according to the classification adopted on Fig. 4.2.

A. Defects in fuel

A.a Defects in pellets

Cracks in pellets are illustrated in Fig. 4.3, whereas Fig. 4.4 shows chips of pellets. On Fig. 4.5 enlarged and broken pellets are shown.

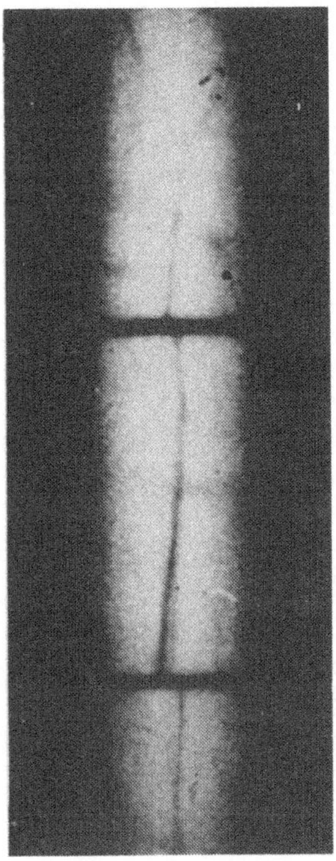

A.a.2
Longitudinal cracks

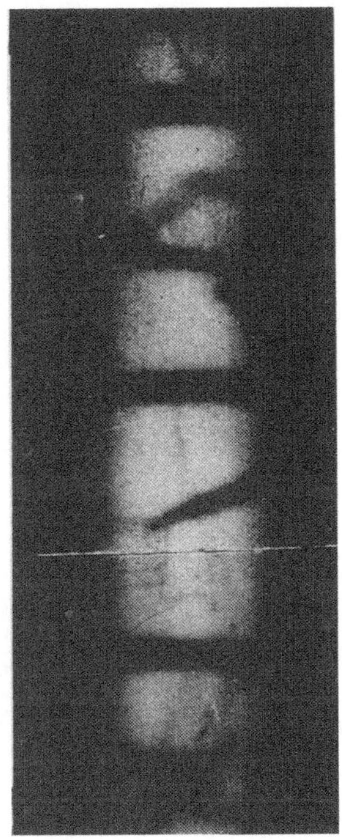

A.a.3
Transverse cracks

Fig. 4.3 Cracks in pellets.

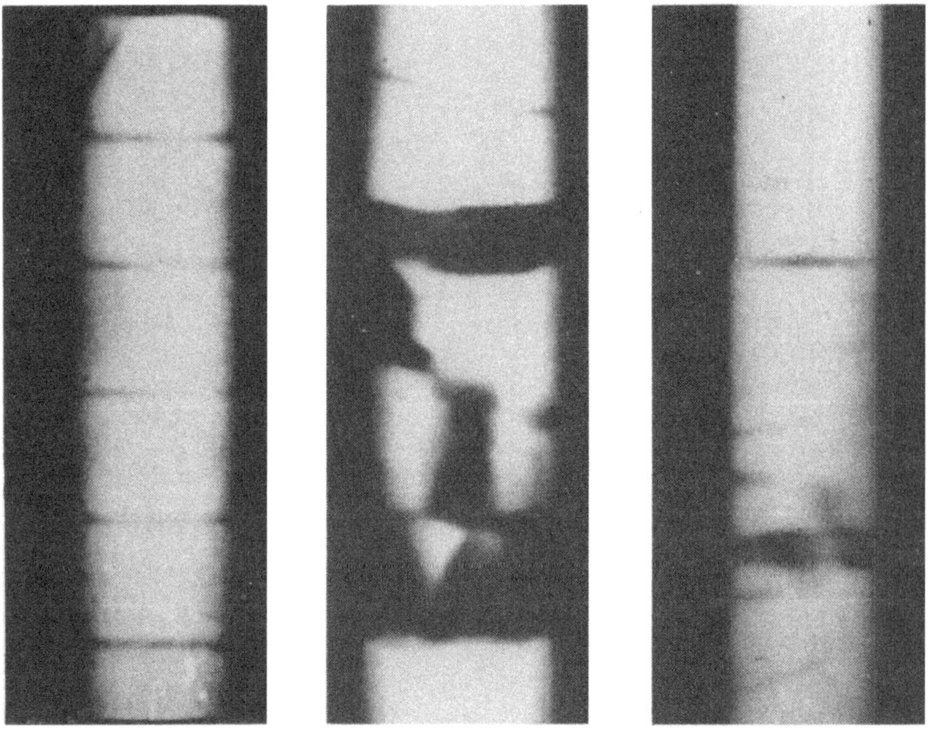

A.a.5
Corner chips

A.a.6
Other chips

A.a.7
Chips in
pellet-to-pellet gap

Fig. 4.4 Chips of pellets

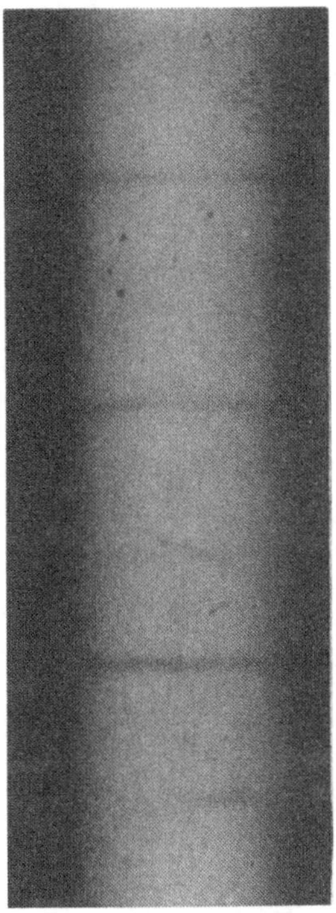

A.a.10
Pellet enlarged

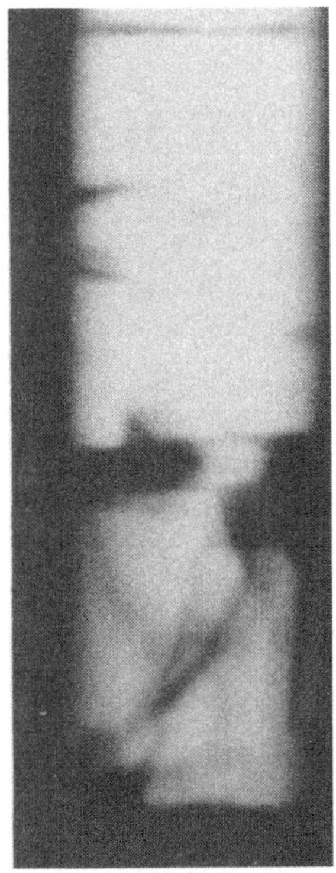

A.a.19
Broken pellet

Fig. 4.5 Enlarged and broken pellets

A.b Defects in pellet-to-pellet gap

On Fig. 4.6 both an enlarged as well as a contracted pellet-to-pellet gap can be seen.

A.b.10
Pellet-to-pellet
gap enlarged

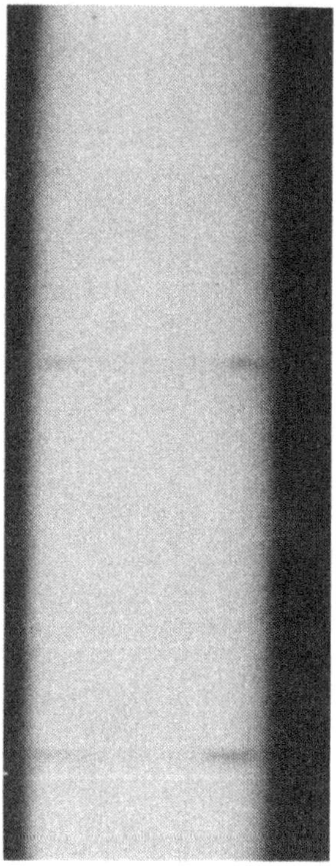

A.b.11
Pellet-to-pellet
gap contracted

Fig. 4.6 Pellet-to-pellet gap enlarged and contracted

A.c Defects in dishing

A filled up and deformed dishing can be seen on Fig. 4.7.

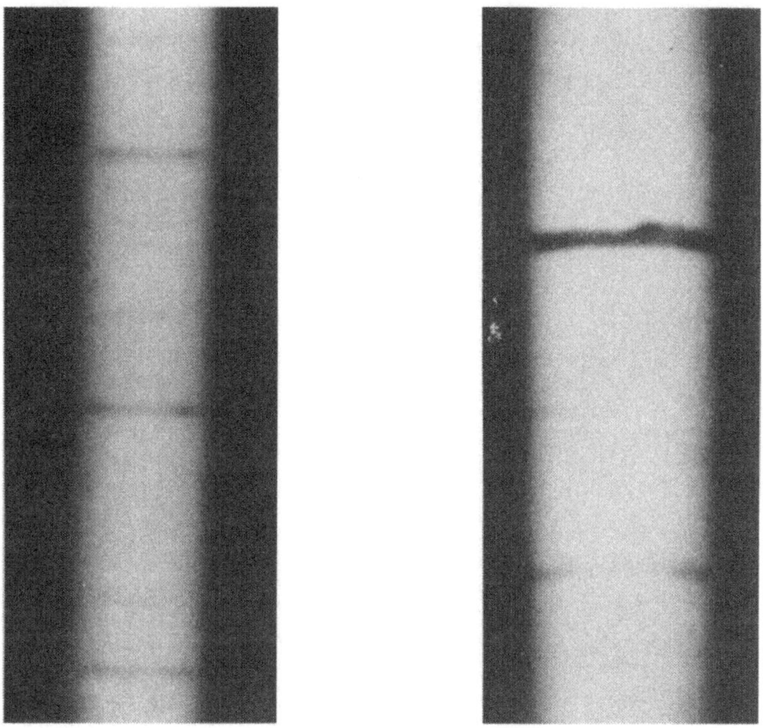

A.c.12
Dishing filled-up

A.c.13
Dishing deformed

Fig. 4.7 Filled-up and deformed dishing.

Central void can be detected in one pellet or going through several pellets (as shown on Fig. 4.8) or can even go through the whole fuel column.

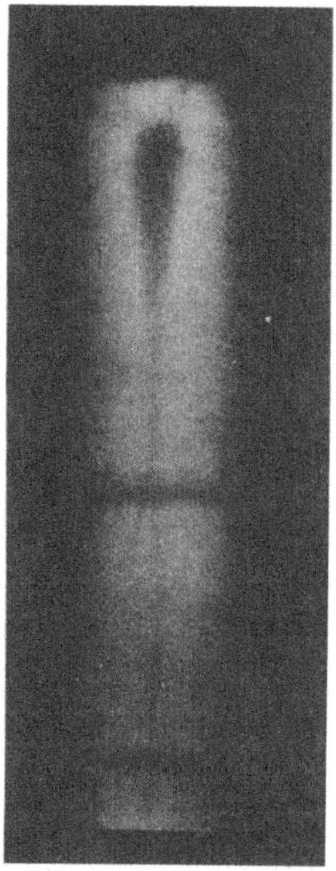

A.d.14
Central void
in one pellet

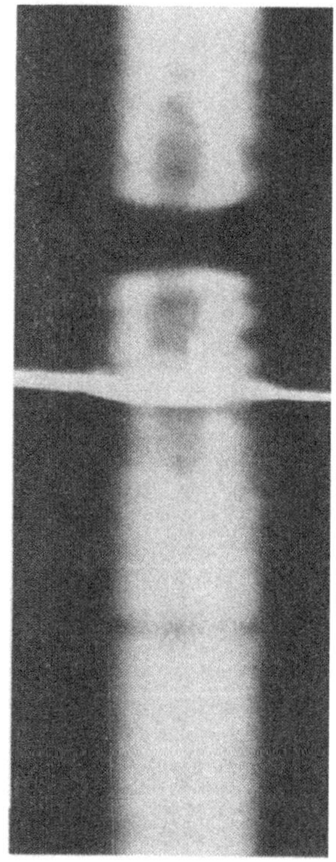

A.d.15
Central void
through several pellets

Fig. 4.8 Central void in one and in several pellets

A.e Defects of fuel-to-clad gap

Defects of fuel-to-clad gap are hard to detect and even harder to reproduce in print. Therefore no such example is given here.

B. Defects in cladding

B.a Deformed and broken cladding

A deformed and broken cladding can be seen on Fig. 4.9.

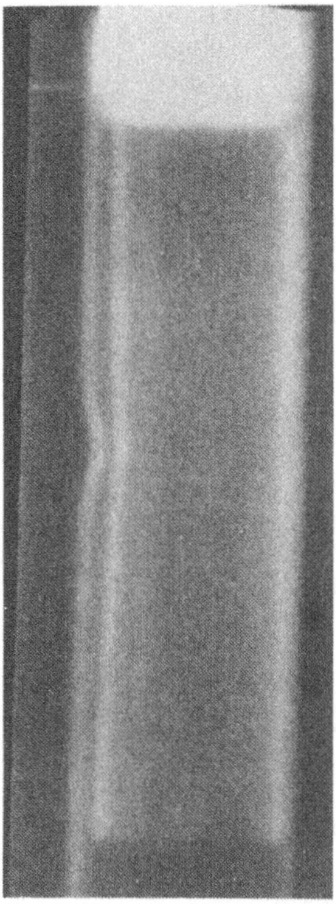

B.a.13
Cladding deformed

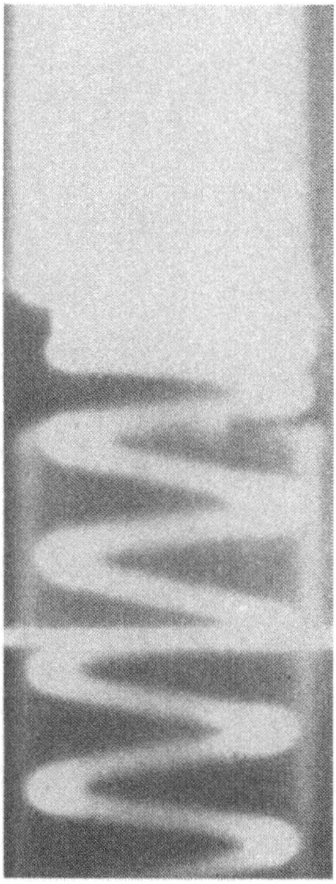

B.a.19
Cladding broken

Fig. 4.9 Deformed and broken cladding

B.a Hydrides in cladding

Hydrides in cladding, although relatively easily detected on neutron radiographs, can hardly be seen when reproduced in print.
Fig. 4.10 shows some hydrides revealed in the cladding.

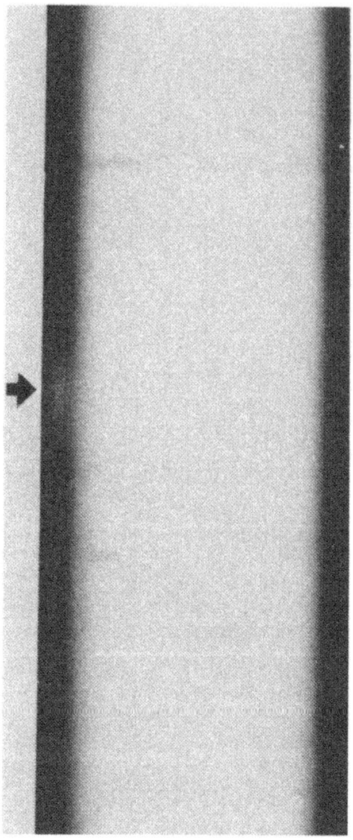

B.a.18
Hydrides in cladding

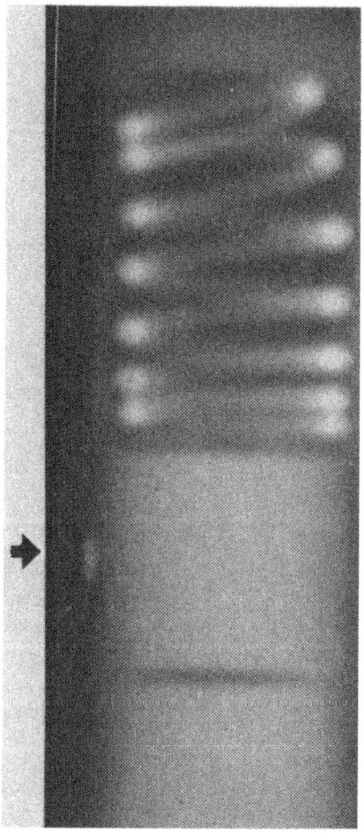

B.a.18
Hydrides in cladding

Fig. 4.10 Hydrides in cladding.

C. Defects in plenum

C.a Defects of spring

Different defects of the spring in plenum are illustrated on Fig. 4.11.

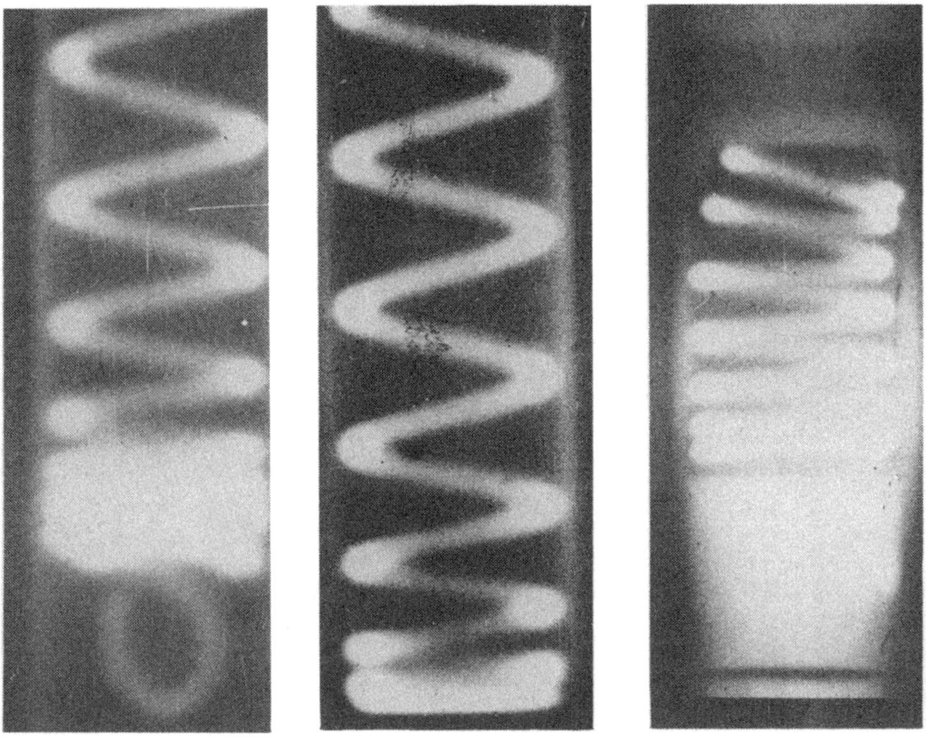

C.a.11
Spring contracted

C.a.13
Spring deformed

C.a.20
Spring dislocated

Fig. 4.11 Defects of the spring in plenum

C.b Defects of spring sleeve

Fig. 4.12 illustrates a broken spring sleeve.

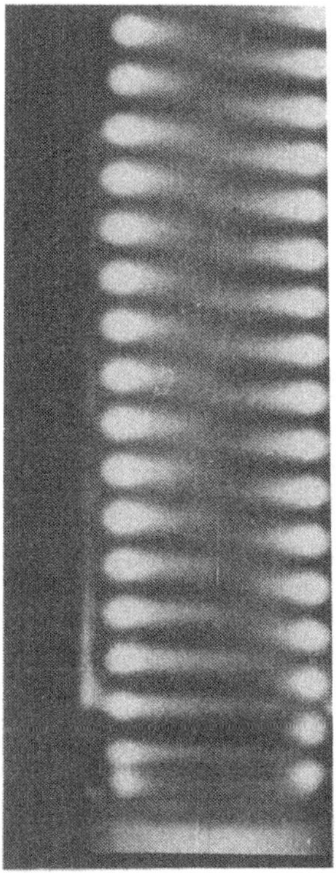

C.a.19
Spring sleeve broken

Fig. 4.12 Broken spring sleeve

C.c Disc

The disc separating the spring of the plenum from the last (or first) pellet can be dislocated, as shown on Fig. 4.13.

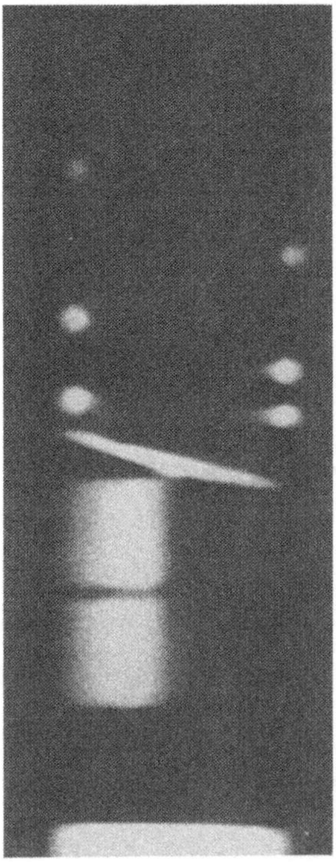

C.c.20
Disc dislocated

Fig. 4.13 Dislocated disc

D. Defects in end plugs

Fig. 4.14 illustrates hydrides detected in the bottom plug. Other defects can be detected by neutron radiography as well.

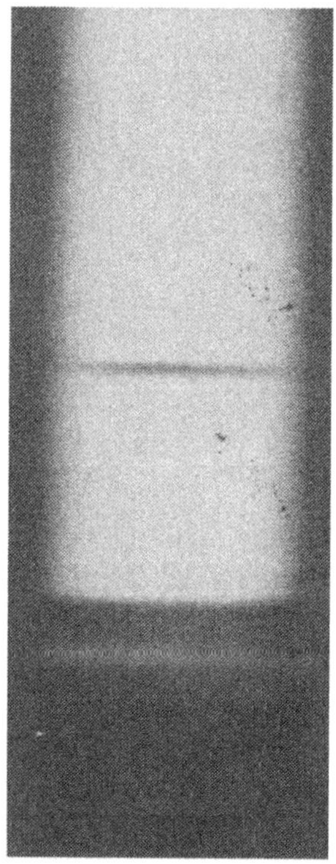

D.a.18
Hydrides in plug

Fig. 4.14 Hydrides in the bottom plug

E. Instrumentation

Defects in various instruments (e.g. thermocouples, pressure transducers) located in fuel pins can be revealed by neutron radiography.
Fig. 4.15 gives an example of a dislocated thermocouple.

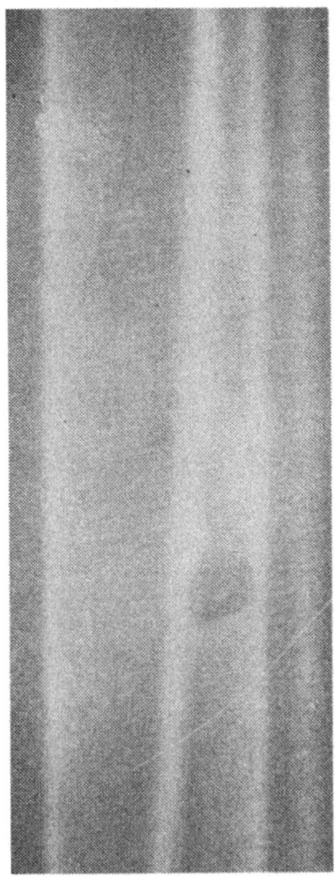

E.a.20
Thermocouple dislocated

Fig. 4.15 Dislocated thermocouple

Defects not shown in the present Collection

In the Atlas only those defects in nuclear fuel are shown which could be chosen from the available neutron radiographs. There are, however, more defects which can most likely occur in nuclear fuel and can be detected by neutron radiography. Those defects were marked "o" on Fig. 4.2.
It is also possible to find some other typical defects in nuclear fuel worth including in this collection. Therefore all persons in possession of such neutron radiographs, missing in this collection, are kindly asked to supply them to :

JRC Petten
Secretary of the NRWG
HFR Division
P.O. Box 2
1755 ZG Petten, The Netherlands

They will be included in the next edition of the Atlas.

4.4 REFERENCES

1. J.C. Domanus :
Neutron Radiographic Findings in Light Water Reactor Fuel.
Metallurgy Department. Risø National Laboratory, June 1979.

Table 5.1 Neutron Radiography Installations in the European Community - Technical Data and Main Utilization.

Site	Facility	Camera type	Collimation Ratio (L/D)	Inlet Diaphr. Dimensions (mm)	Collimator Lining	Beam Dimensions (mm), at obj. plane	Thermal Neutron Fluence rate, at obj. plane ($m^{-2}s^{-1}$)	Max. Obj. Dimensions (mm)	Min. Distance Object/ Image Plan (mm)	Geometr. Unsharpness	Cd-Ratio	Other Spectrum Inform.	Utilization					Approx. Number of Exposures per year
														nuclear				
													non-nuclear	LWR fuel pins	LMFBR fuel pins	Irradiation devices	Miscellan.	
Cadarache	LDAC	dry	13,5	70 x 30	Cd	500 x 100	ca. 10^9	500 x 100 x 60	ca. 1	--	ca. 9				X			1200 (since Aug. 1970)
Casaccia[1])	TRIGA-RC 1	dry	50	Ø = 48	borated paraffine L = 2200 mm $Ø_{min}$ = 48 mm (conical tube)	Ø = 120	5 . 10^{11}	520 x 27 Ø	8	0,15 mm	1,65 (measured with Au)		--	X	X	--	--	
Fontenay-aux-Roses	TRITON	dry	180 canal axial				7 . 10^{10}	180x240[2])	2	below visual percept. faculty			X	X	X	X	--	2500
			110 to 760 canal latér.				4 . 10^{10}	300x400[2])	2				X	X	X	X	--	
Geest-hacht	FRG 1	dry	375	20	B_4C	Ø = 180	5.10^{10} [3])	300 x 300		26 - 50 µm	100 (Dy 0,1 mm)		X	--	--	--	X	50 to 100
	FRG 2	pool	100	Ø = 20 (other possible)	Sandwich: boral, iridium, Cd	100 x 400	10^{11}	100 x 100 length 1700	10	0,1-0,5 mm	5 (Au 25µm) 90 (Dy 0,1 mm)		--	X	--	X	--	50
	1kCi Sb-Be neutron source	dry	10 - 20	20	no collimator	200x400	1 - 5. 10^8	3000x1000	1	--	2 (Au 25µm)		--	--	--	--	X	20 to 50
Grenoble	MELU-SINE	dry	125 and 390	Ø = 50 and 16,2	B_4C + In	Ø = 400	2 . 10^{11} and 2 . 10^{10}	normal length < 2000 (possible modificat. for bigger objects)	2	ca. film grain un-sharp-ness	7,4	- Collimator nose in D_2O tank - Cold neutr. beam facility with Be-filter	X	--	--	--	X	100 to 200
	SILOE	pool	380	Ø = 6	first 200 mm: Cd,In,Dy, Au,Gd general lining: B_4C,In	400 x 132	8,5.10^{11}	140 x 140 (section)	6	"	2,4	ϕ_{epi} (> 0.52 eV) 7,9 . 10^{10} $m^{-2}s^{-1}$	--	X	X	X	--	100 to 150

1) not operational at present 2) max. dimensions of usable film 3) $\phi_f = 10^{11}\ m^{-2}s^{-1}$

Table 5.1 Contd.

Site	Facility	Camera type	Collimation Ratio (L/D)	Inlet Diaphr. Dimensions (mm)	Collimator Lining	Beam Dimensions (mm), at obj. plane	Thermal Neutron Fluence rate, at obj. plane ($m^{-2}s^{-1}$)	Max. Obj. Dimensions (mm)	Min. Distance Object/ Image Plan (mm)	Geometr. Unsharpness	Cd-Ratio	Other Spectrum Inform.	Utilization: non-nuclear	Utilization: nuclear: LWR fuel pins	Utilization: nuclear: LMFBR fuel pins	Utilization: nuclear: Irradiation devices	Utilization: nuclear: Miscellan.	Approx. Number of Exposures per year
Harwell	DIDO (Beam 6H)	dry	50 (5m station)	Ø = 150	Boral	Ø = 150	10^{11}	500 x 500 x 500	0	∞	0	beryllium filtered beam	X	--	--	--	--	mainly dynamic imaging and recording
			300 (25m station)	Ø = 150	Boral	Ø = 500	3×10^9	h. 1600 l. 1700 w. 3400	0	∞	0	"	X	--	--	--	--	"
	(Beam 6HGR9)	dry	50	Ø = 19	Cd	Ø = 180	8×10^{11}	diam. 260 l. 1730	0 to 150		0		X	X	X	X	X	500 excl. routine radiographs of turbine blades
Karlsruhe	FR 2	dry	46 to 185	$81 cm^2$ to $5{,}07 cm^2$	1 mm Cd	250 x 170	0,5 to 4,5 x 10^{10}	diam. 135 l. 6000	direct contact			neutrons from therm. column	4)	--	--	X	--	190 in 1978
Mol	BR 1	dry	75	Ø = 30	Pb; Boral	300 x 300	$1{,}1 \cdot 10^{10}$	3000x200 x 40	1	15 μm (min.)	> 50 (for Au)			non destructive quality control of plutonium for LWR and fast reactors				250
	BR 2	pool	240 typical	Ø = 11	Boral (B_4C)	100 x 600	$3 \cdot 10^{11}$	w. 100 l. 3000	0 (min.) 28 (typ.)	100 μm (typical)	40	spectrum varies with reactor loading	--	X	--	X	--	150
Petten JRC	HFR (PSF)	pool	237	8	1/4'' B_4C	600 x 80	$2{,}3 \cdot 10^{11}$	150 x 200 x 1560	0,5	2 μm (without object)	10,2		--	X	X	X	X	300
	(HB8)	dry	500	Ø = 8	B_4C	160 x 100	10^{11}	w. 100 l. 4500	2	4 μm	not yet measured		--	X	X	X	--	new installation
Petten ECN	LFR	dry	127	Ø = 15 (other possible)	--	Ø = 250	$3 \cdot 10^9$	7500x5000	in contact	8 μm (without object)	40,4 (with manganese foils)	$n/\gamma = 1{,}6 \cdot 10^6$	X	--	--	X	X	100

4) only for demonstration

Table 5.1 Contd.

Site	Facility	Camera type	Collimation Ratio (L/D)	Inlet Diaphr. Dimensions (mm)	Collimator Lining	Beam Dimensions (mm), at obj. plane	Thermal Neutron Fluence rate, at obj. plane ($m^{-2}s^{-1}$)	Max. Obj. Dimensions (mm)	Min. Distance Object/ Image Plan (mm)	Geometr. Unsharpness	Cd-Ratio	Other Spectrum Inform.	Utilization					Approx. Number of Exposures per year
													non-nuclear	nuclear				
														LWR fuel pins	LMFBR fuel pins	Irradiation devices	Miscellan.	
Risø	DR 1	dry	110 in vertical, 27,5 in horizontal direction	20 vertically, 80 horizontally	graphite	twice 100 x 100	1,8.10^{10} (left part) 1,4.10^{10} (right)	twice 100 x 100 to be radiographed, otherwise no dimens. limits	in contact	$\frac{20d}{2200-d}$ [4] (vertic.) $\frac{80d}{2200-d}$	4,2 (left port) 3,8 (right) Au	25 R/h gamma at object for open beam port	X	X	--	--	X	500
Saclay	OSIRIS	pool	148	16 x 16	Boral 8 mm	150 x 600	6,5.10^{11}	l. < 2500	> 12		3,83		--	--	--	X	--	100 to 130
	ISIS	pool	137	16 x 16	Boral thickness 8 mm + (In,Cd on 200 mm)	150 x 600	0,3 to 1,3 x 10^{11}	l. < 1800	> 18		5,64 to 2,54		device is used if OSIRIS is not available					40 to 50
		dry	94	Ø = 40	boron powder 10 to 12 mm	100 x 150	8,4.10^{10}	l. < 4000 two tubes Ø = 20, ; or one tube Ø = 43	1 to 7		2,44		--	X	X	--	--	250 to 300
	ORPHEE	dry	divergent: 15'	neutron guide		150 x 25	5.10^{12} [6]	300x400 [7] 150x1010	1	not yet determined	0	sub-thermal	X	--	--	X [8]	--	first tests in May 1981 [9]
Valduc	MIRENE	dry	100	20 x 30 tangent beam	1934 mm	180 x 240	2,6 x 10^{12} [10]	large dimens. > 2m	3500		9		X	-- [11]	-- [11]	X	X	
				30 x 30 axial beam	1722 mm	300 x 300	8,9 x 10^{12} [10]	"	"		5.9		X	-- [11]	-- [11]	X	X	

5) d = object thickness in mm
6) expected
7) film dimensions
8) before irradiation
9) The installation at ORPHEE will replace the TRITON installation.
10) m^{-2}/pulse instead of $m^{-2}.s^{-1}$
11) possible.

Table 5.2 Neutron Radiography Installations in the European Community - Exposure Techniques.

Site	Facility	Converters Used	Films Used	Track Etch Film Used	Typical Expos. Times	Beam Purity and/or Image Quality Indicators Used	Film Development Procedure (bath, temp., time)	Special Dark Room Equipment
Cadarache	LDAC	Dy or In	KODAK - Industrex	--	2 min. (total time of neutron volley)	no	LX24, 5 min. at 20 °C	Négatoscope
Casaccia	TRIGA RC1	In	KODAK - Kodirex	no	10 - 20 s.	VISQI	LX24, room temperature	no
Fontenay-aux-Roses	Triton	Gadolinium 250 μm and 25 μm	KODAK - Industrex A, M, R, mono-couche	CN80-15	2 - 20 min.	I.Q.I.	Manual development in vertical troughs. Revelator SOPRECO: 20 min; Fixator rapid ILFORD	1)
Geesthacht	FRG 1	Dy 0,1 mm In 0,1 mm	Structurix D4, D2	CA80-15 CA80-15B	20 - 60 min.	VISQI		
	FRG 2	Dy 0,1 mm	Structurix D4	--	15 min.	--	20 °C, 5 min.	
	1kCi Sb-Be neutron source	Dy 0,1 mm	Structurix D7	CA80-15B	6 h	--	etching 6h NaOH 50 °C, 30 min.	
Grenoble	MELUSINE	Gd 25 μm	KODAK - MX, M, R single coated	CN80-15	2 min. or 20 min. with cold neutrons (film M)	Research Chemicals VISQI Test Object	Fixator Kodak AL4, 10 min. at 20 °C, with thermostatic control. Revelator Kodak LX24, 5 min. at 20 °C, with thermostatic controls.	Enlarger, Contact Reproduction, Light-Box, Profile Projector, Densitometer and Micro-densitometer.
	SILOE	Dy 50 μm and 100 μm	KODAK - M and R single coated	CN80-15	6 min. (film R)	no	"	"
Harwell	DIDO (Beam 6HGR9)	Gd and In foils. Direct and transfer	KODAK - Industrex C and S.R. ILFORD - SP 352 line film	KODAK CA80-15B	direct : 3 - 60 s. indirect : 5 - 15 min.	not used to date	Standard developer 20 °C, 4 min.	Densitometer
	DIDO (Beam 6H)	2)						

1) Dark room facilities available :

- One laboratory installed within the reactor hall, near the neutron-radiography installations, for treatment and duplication of silver-base films.
- One laboratory outside the reactor, for treatment of nitrate/cellulose -based films and reproduction on photographic paper etc.

2) Exposure techniques, converters used :

Real time dynamic imaging using screens, image intensifier, T.V. Camera and video recorder.

Table 5.2 Contd.

Site	Facility	Converters Used	Films Used	Track Etch Film Used	Typical Expos. Times	Beam Purity and/or Image Quality Indicators Used	Film Development Procedure (bath, temp., time)	Special Dark Room Equipment
Karlsruhe	FR 2	Indirect, with Dy foils	AGFA D2, D4, D7 Osray	KODAK CA80-15	1 h neutron/foil 1 h foil/D7 film	not used	Developer AGFA G 150 room temp., 10 min.	no special equipment
Mol	BR 1	25 μm Gd	Structurix D4 (Agfa-Gevaert)	--	15 - 30 min	reference fuel pin	procedure as given in AGFA-Gevaert manuals)	--
	BR 2	Dy 0,1 mm In 0,1 mm	Structurix D2, D4 (Agfa-Gevaert)	--	6 - 8 min.	VISQI	AGFA G 150, 20 $^{\circ}$C, 5 min.	
Petten	HFR (PSF)	Dy 0,1 mm Kodak BNI	KODAK M and SR	KODAK-CN85	16 min. for Dy 7 min. for CN85	no	3)	13 x 18 cm Enlarger
	HFR (HB8)	Kodak BNI and 93 %enr. ^{10}B	no	CA80-15	5 min.	no	etching time 30 min. in NaOH 100g/L at 46 $^{\circ}$C	13 x 18 cm Enlarger
	LFR	Gd, 100 μm	Agfa - D7 KODAK -SR	Yes	D7 : 16 min. SR: 120 min.	no	Standard procedure	MacBeth Densitometer
Risø	DR 1	direct : Gd 50 μm transfer : Dy 100 μm	Agfa-Gevaert Structurix D4 KODAK - Industrex SR	KODAK - Pathé CA80-15B CN85-IB CN85	30 min. for D4 90 min. for track-etch	ASTM E545-75	X-ray film: 20 $^{\circ}$C hand processing 4 min.	X-ray Film Producing Tanks and Thermostatic Etches
Saclay	OSIRIS	Dy	KODAK - monolayer (SR 54)	--	15 min.		classic methods for x-ray films	Négatoscope, Profile Projector, Contact Reproduction, Enlarger, Densitometer, Polaroid Screens, X-ray Film Process, Thermostatic Etch Bath
	ISIS	transfer : In or Dy	KODAK - Industrex M or SR 54	--	20 min.		"	
	ISIS	direct : n,α ^{10}B or ^{6}LiF	KODAK CN85	--	3 - 10 min.		"	
		transfer: Dy	KODAK SR 54					
Valduc	MIRENE	Gd, Dy, In	KODAK type A KODAK type M	Yes	3 min.	IQI, BPI	Manual, LX24, 5 - 8 min. at 20 $^{\circ}$C	Yes

3) a) If necessary :

- foil (Dy-0,1 mm) exposure time : 16 min.
- transport time of the activated foil to the dark room : 10 min.
- foil transfer on Kodak M film : 35 min.
- hereafter transfer on Kodak SR film : overnight

b) Film development :

- 5 min. at 20 $^{\circ}$C in Kodak DX80
- rinsing for 2 min. in running water
- fixing for 4 min. in Agfa G334
- washing for 20 min. in running water

c) Track-etch film :

- exposure of Kodak CN85 for 7 min.
- etching time 30 min. in NaOH (100 g/L) at 46 $^{\circ}$C
- all track-etch films are copied on Kodalith 2571.

Table 5.3 Neutron Radiography Installations in the European Community. Future Needs and Requirements.

Site	Facility	Qualitative Analysis			Quantitative Analysis			
		Standards Used	In-House Atlas	Other	Standards Used	Profile Projector	Microdensitometer	Other
Cadarache	LDAC	unirradiated fuel pins	--	--	--	--	--	--
Casaccia	TRIGA RC1	--	--	--	--	--	Yes	--
Fontenay-aux-Roses	TRITON		1)			2)		
Geesthacht	FRG 1	--	--	--	homemade steps of absorbers (lucite, Gd-foils)	--	fast photometer G III Jenoptik Jena	--
	FRG 2	unirradiated fuel and absorber pins, dummy rigs	--	--	dummy for dimensions	--	"	--
	1kCi Sb-Be neutron source	--	--	--	--	--	--	--
Grenoble	MELUSINE	--	Yes	--	--	Scanning Profile Projector	Densitometer and Scanning-Micro-densitometer	--
	SILOE	--	Yes	--	pre-irradiation picture	"	"	--
Harwell	DIDO (Beam 6H)	not relevant, work too disparate			not relevant	not relevant	not relevant	--
	DIDO (Beam 6HGR9)	no general standards, scientific examination	None	--	--	--	Yes	--

1) Qualitative Analysis :

In most cases, this examination is done by the client. Technical assistance by the S.E.T. is offered only when specifically asked for. Image quality can be guaranteed however (well-focused photos without handling traces, chemical pollution, etc.)

2) Quantitative Analysis :

- Dimensional measurements on film
- Measurements to determine the materials' homogeneity
- Use of optical measurement installation, densitometer and micro-densitometer.

Table 5.3 Contd.

Site	Facility	Qualitative Analysis			Quantitative Analysis			
		Standards Used	In-House Atlas	Other	Standards Used	Profile Projector	Microdensitometer	Other
Karlsruhe	FR 2	not in use			--	--	Joyce-Loebl LTD with Auto-densidater	--
Mol	BR 1	reference fuel pin containing pellets with different enrich-ment and Pu grains of different sizes	--	--	reference fuel pin containing pellets with different enrich-ment and Pu grains of different sizes	--	--	--
	BR 2	--	--	--	optical micro-meter	--	fast photometer VEB Carl Zeiss Jena	--
Petten	HFR (PSF 10)	--	no	no	--	Nikon 6 CT2	no	no
	HFR (HB8)	--	no	no	--	,,	no	no
	LFR	--	no	no	--	,,	no	no
Risø	DR 1	ASTM E 545 - 75	classification of defects revealed by neutron radiography	--	ASTM E 545 - 75 Calibration fuel pin	Nikon 6C (10x magn.)	Baird double beam densitometer	Special Cd device for L/d measure-ments
Saclay	ORPHEE 3)							
	OSIRIS ISIS	homemade dummy rigs and un-irradiated fuel and absorber pins	--	--	--	Orama 500 (10x, 20x, 50x)	MacBeth quanta log	
Valduc	MIRENE	--	--	--	--	--	--	Densitometer

3) Neutron radiography installation is being tested at present.

Table 5.4 : Neutron Radiography Installations in the European Communities. Future Needs and Requirements.

Needs and Requirements in the field of Research and Development	Ge	- contrast enhancement of images on track etch films
	K, P, R	- track etch technique (improvements)
	K	- copying nitrocellulose films
	F, S	- reproducibility (density, image quality)
	H	- image quality
	F, P, S	- reduction of inherent scattering in numerous materials (neutron energy, anti-scatter grids)
	H, Gr	- dynamic imaging
	Ge	- converters of higher sensitivity
	Gr, K, M, P, R	- technique of dimensional measure-
	M	- ments
	M	- epithermal neutron radiography
	Ca, M, Gr	- tomography
	P	- biomedical application
	C	- general
Needs and Requirements for Practice Guide	Gr, H	- track etch
	R	- classification and collection of defects revealed by neutron radiography
	R	- recommended procedures for direct and transfer methods
Needs and Requirements for Standards	Ce	- dimension measurements indicator for a) resolution b) parallaxis c) magnification
	M	- development of a universal reference fuel pin
	P	- Risø calibration pin
	R	- calibration standard for dimension measurements
	P, R, F, S	- indicator for a) beam quality b) image quality
	Gr, M	- general yes (2x)
	F, S	- standard procedures for control and for a Non-Destructive Control Manual

C - Casaccia
F - Fontenay-aux-Roses
Ge - Geesthacht
Gr - Grenoble
H - Harwell
K - Karlsruhe
M - Mol
P - Petten
R - Risø
S - Saclay
V - Valduc
Ca - Cadarache

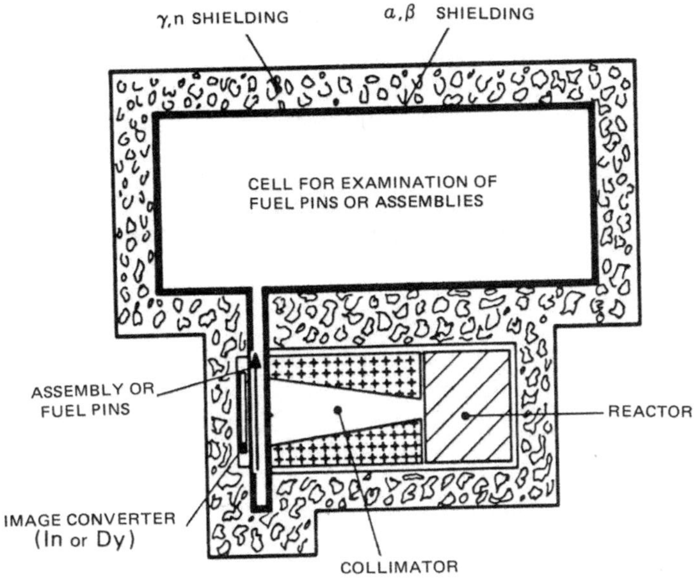

REACTOR INSTALLATION SCHEME (LDAC AND CEI)

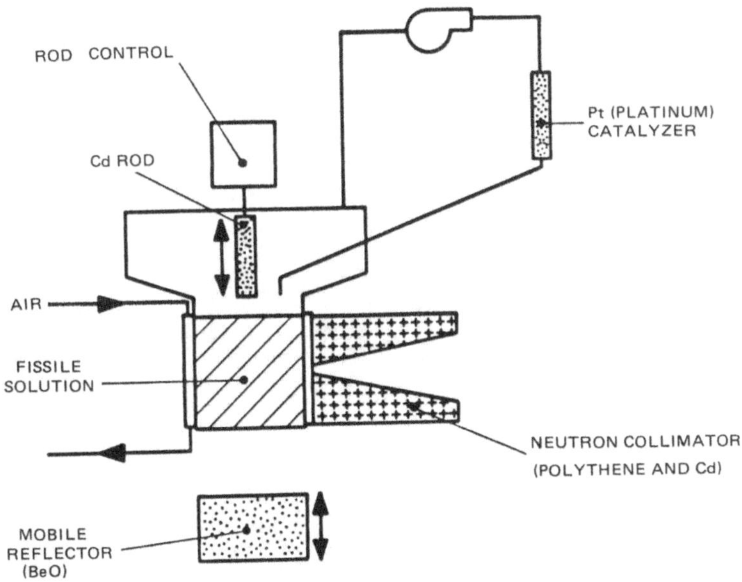

SCHEME OF LDAC REACTOR (RAPSODIE)

Fig. 5.1 Neutron Radiography Facilities at CEN Cadarache.

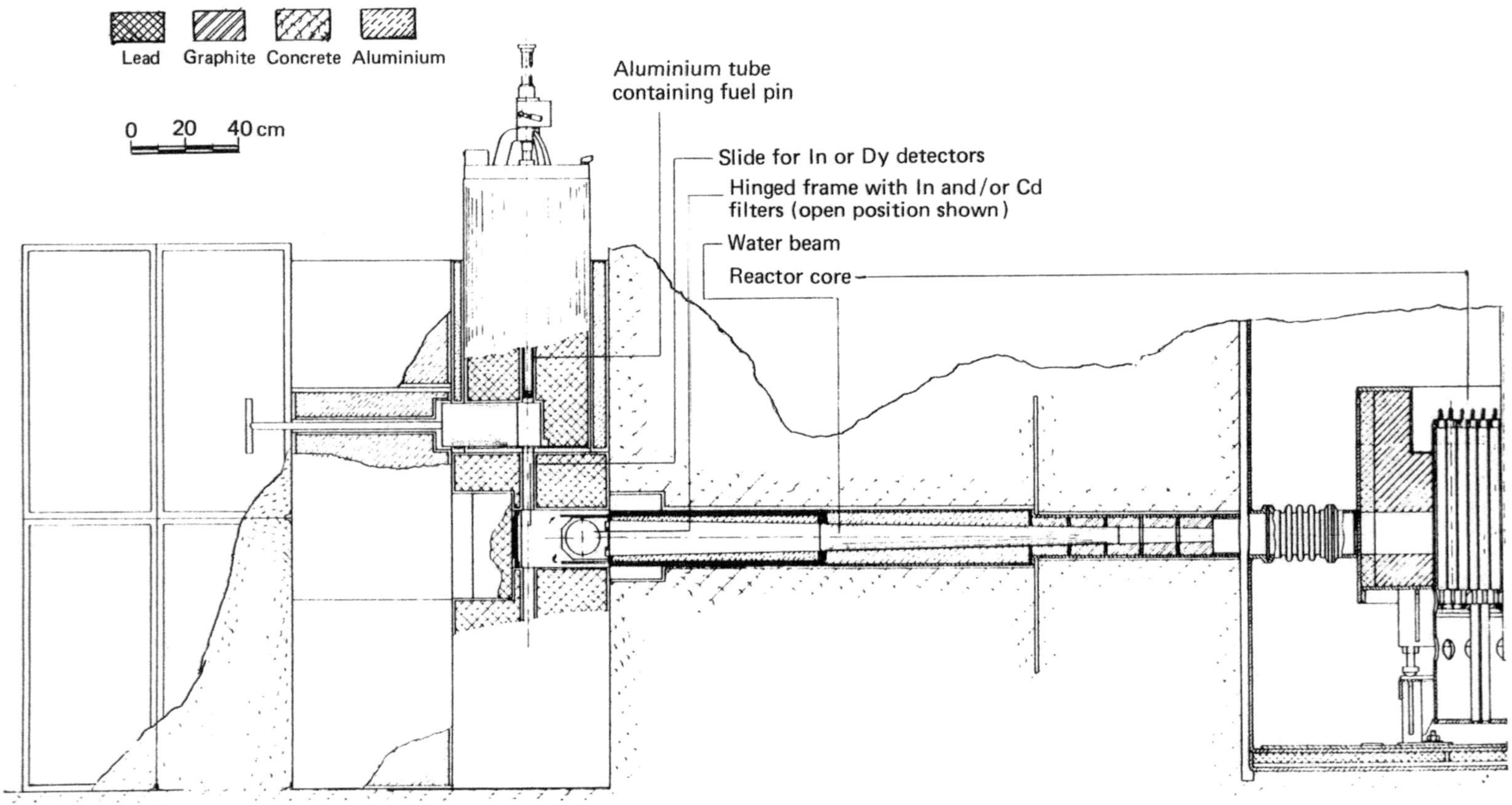

Fig. 5.2 Sketch of the Neutron Radiography Facility at CNEN-CSN, Casaccia.

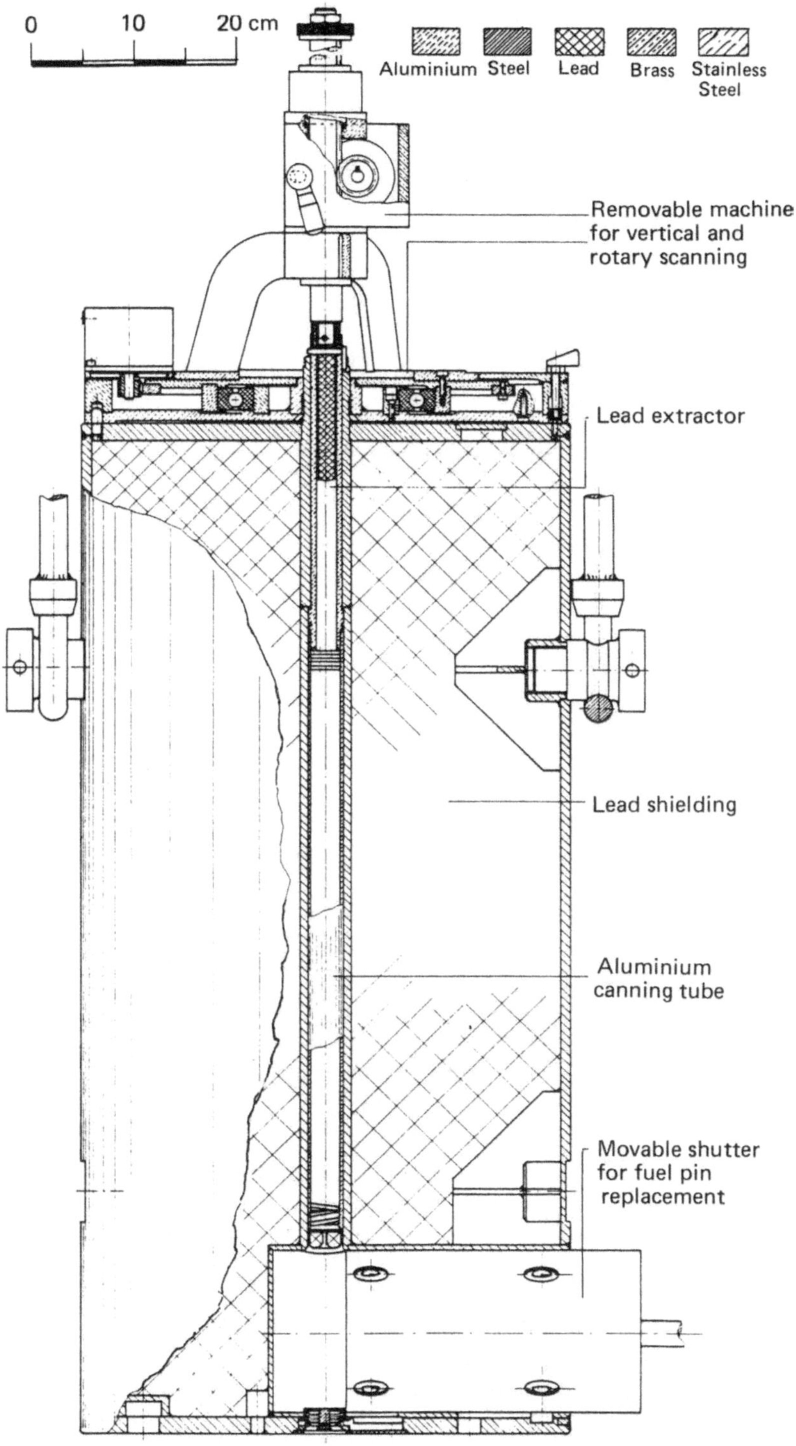

Fig. 5.3 Neutron Radiography Installation at CNEN-CSN Casaccia - Dual Purpose Transport/Exposure Container.

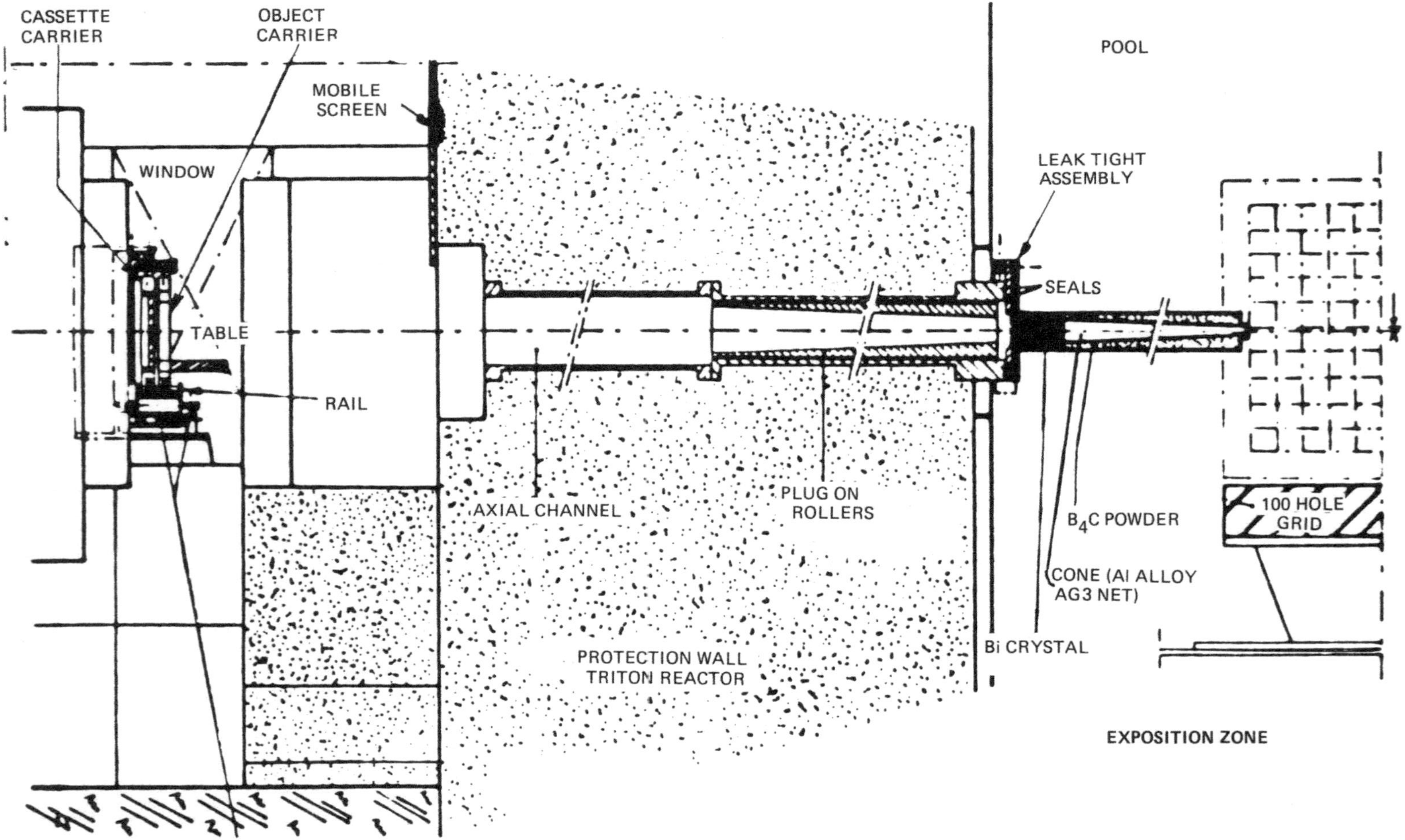

Fig. 5.4 Neutron Radiography Installation at the Axial Channel of the TRITON Reactor, Fontenay-aux-Roses.

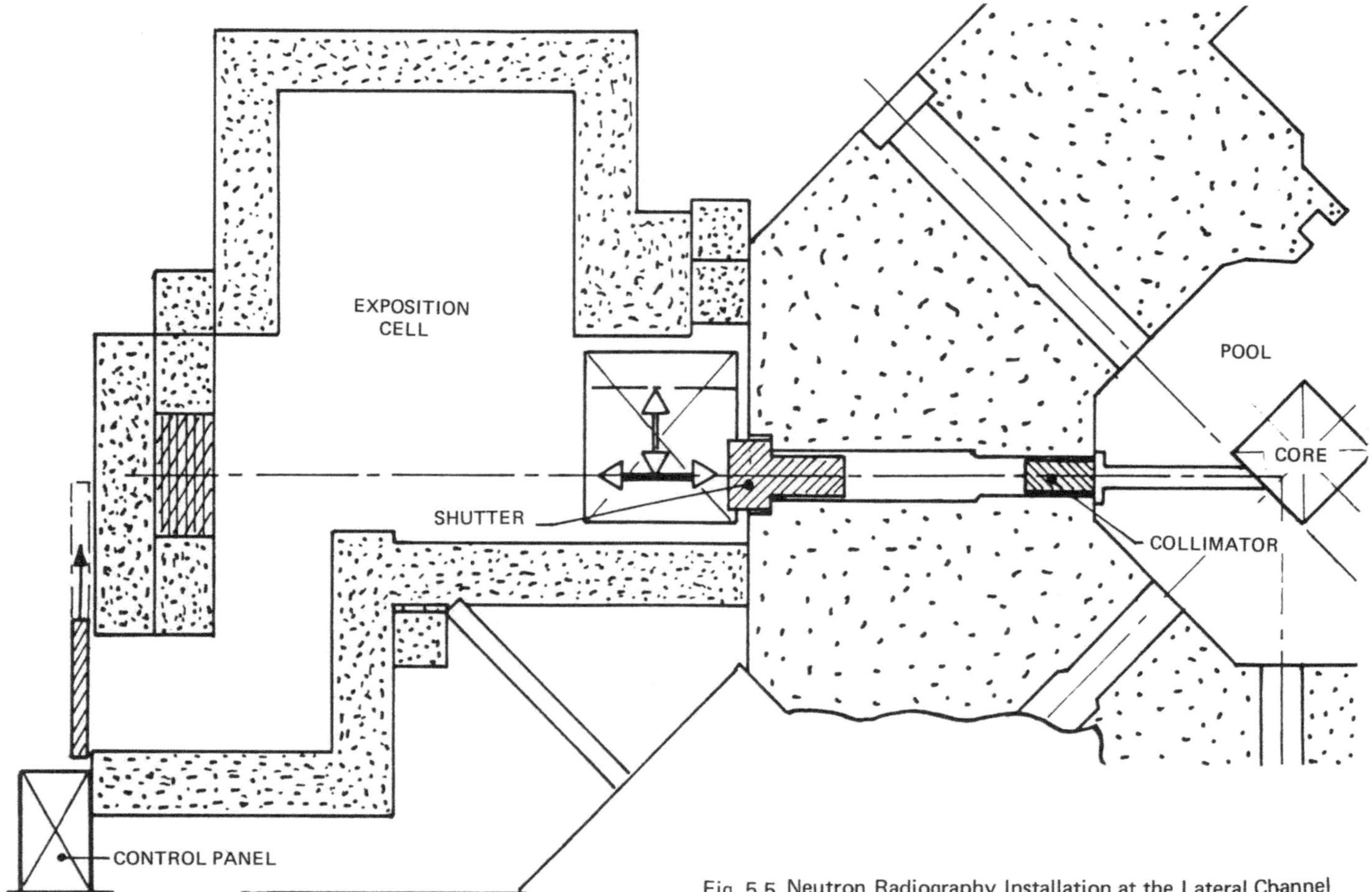

Fig. 5.5 Neutron Radiography Installation at the Lateral Channel of the TRITON reactor, Fontenay-aux-Roses.

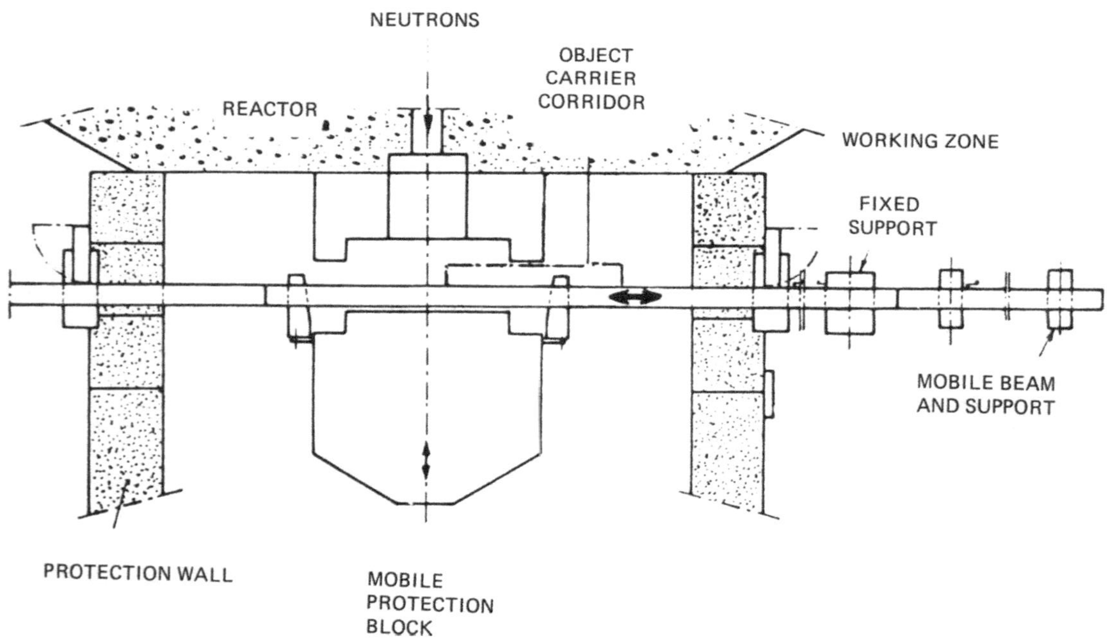

Fig. 5.6 Utilization Zone of the Neutron Radiography Installation at the TRITON Reactor, Fontenay-aux-Roses.

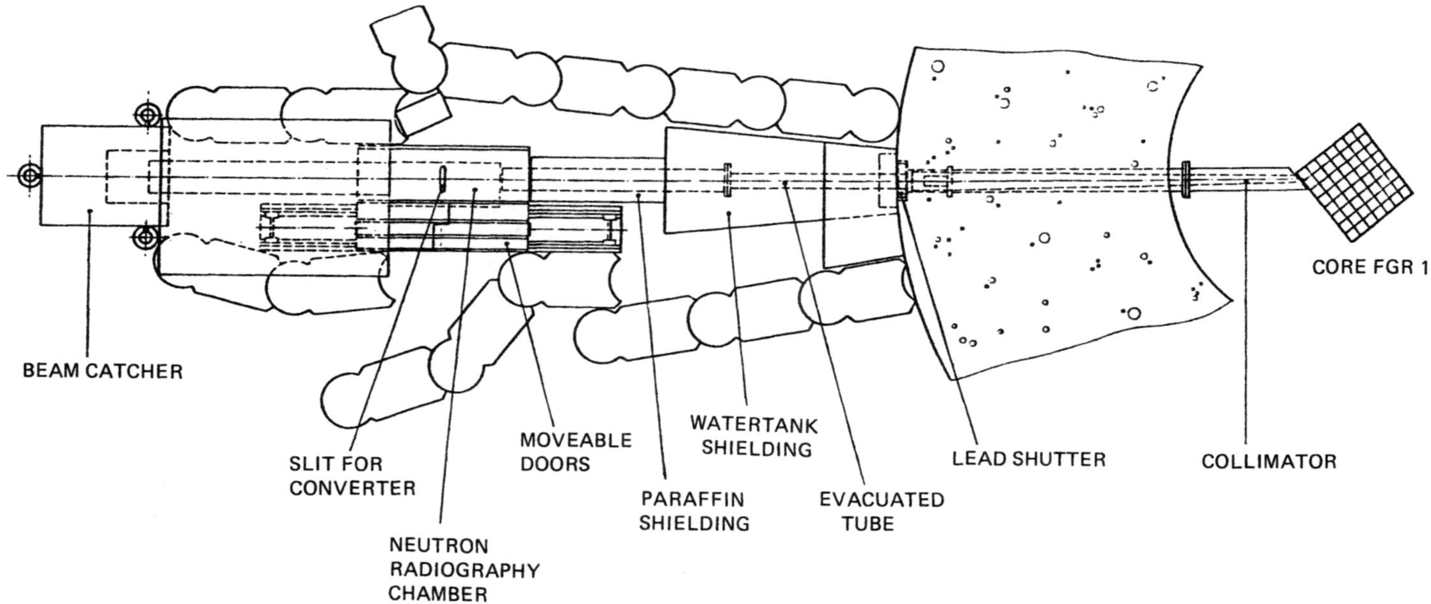

Fig. 5.7 Beam hole neutron radiography facility GENRA I at the FRG 1 reactor, GKSS Geesthacht.

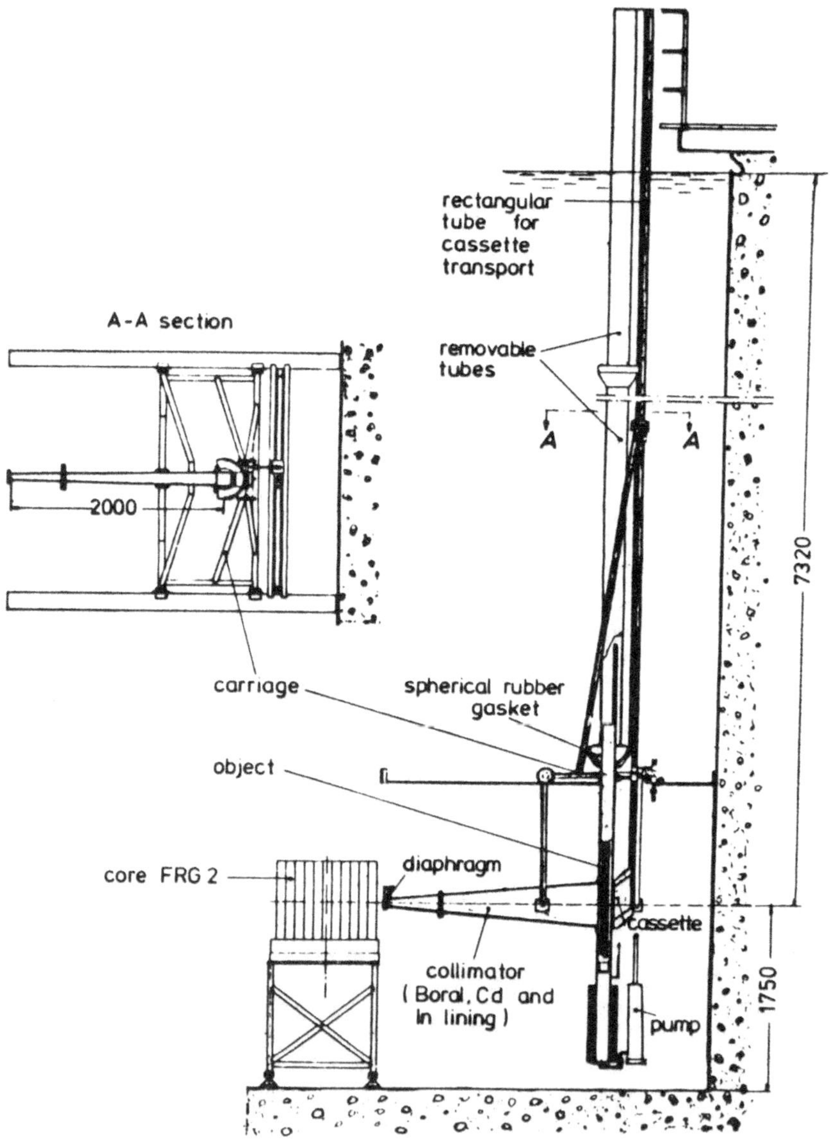

Fig. 5.8 Neutron radiography underwater facility GENRA II at the FRG2 reactor, GKSS Geesthacht.

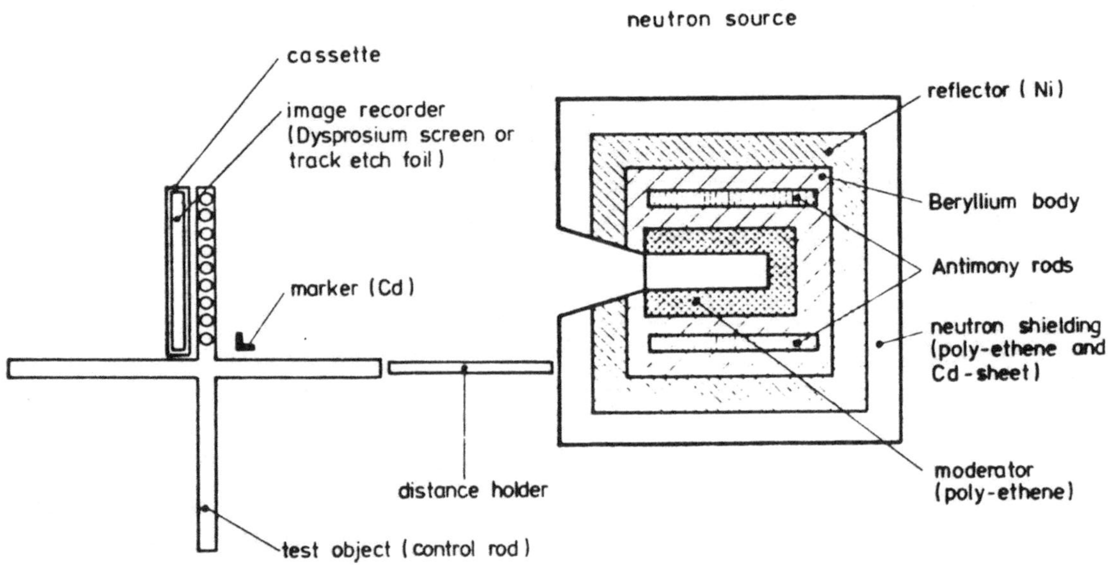

Fig. 5.9 Arrangement for neutron radiography of a control rod in a hot cell, with antimony-beryllium neutron source at GKSS Geesthacht.

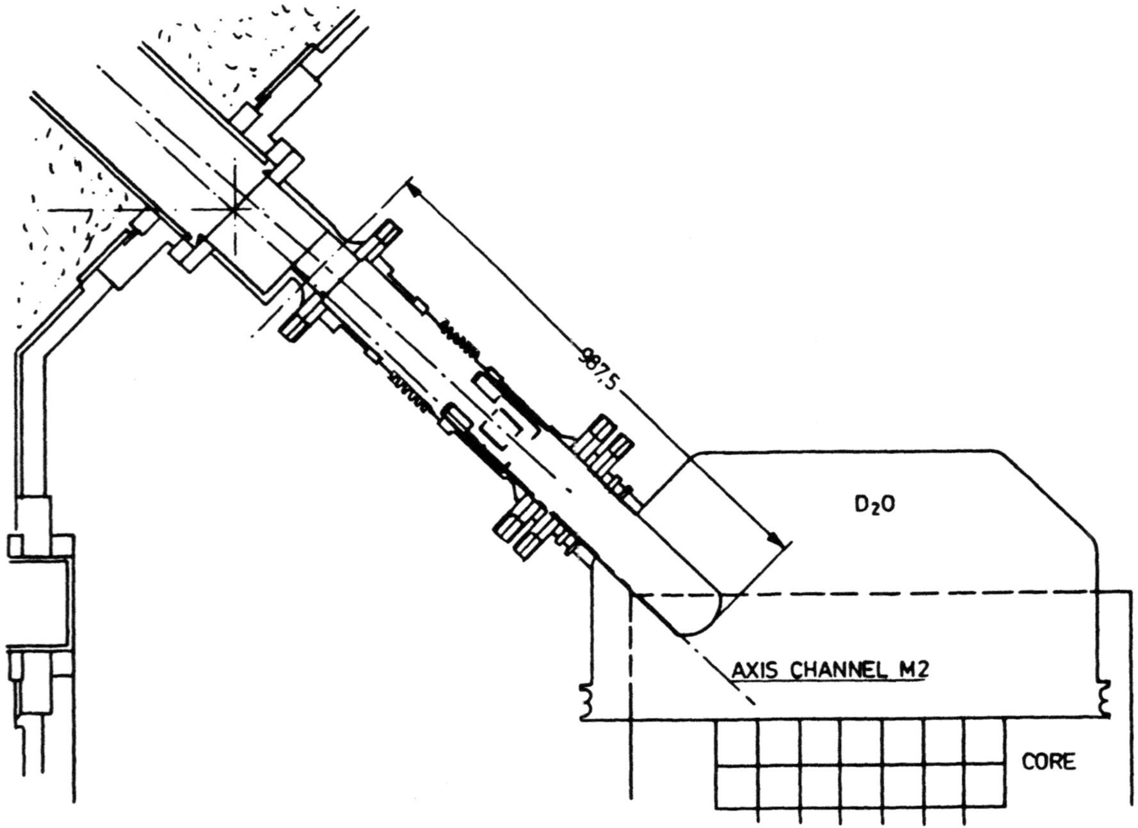

Fig. 5.10 Neutron Radiography Installation at the MELUSINE Reactor, Grenoble. Schematic Drawing of the Neutron Beam.

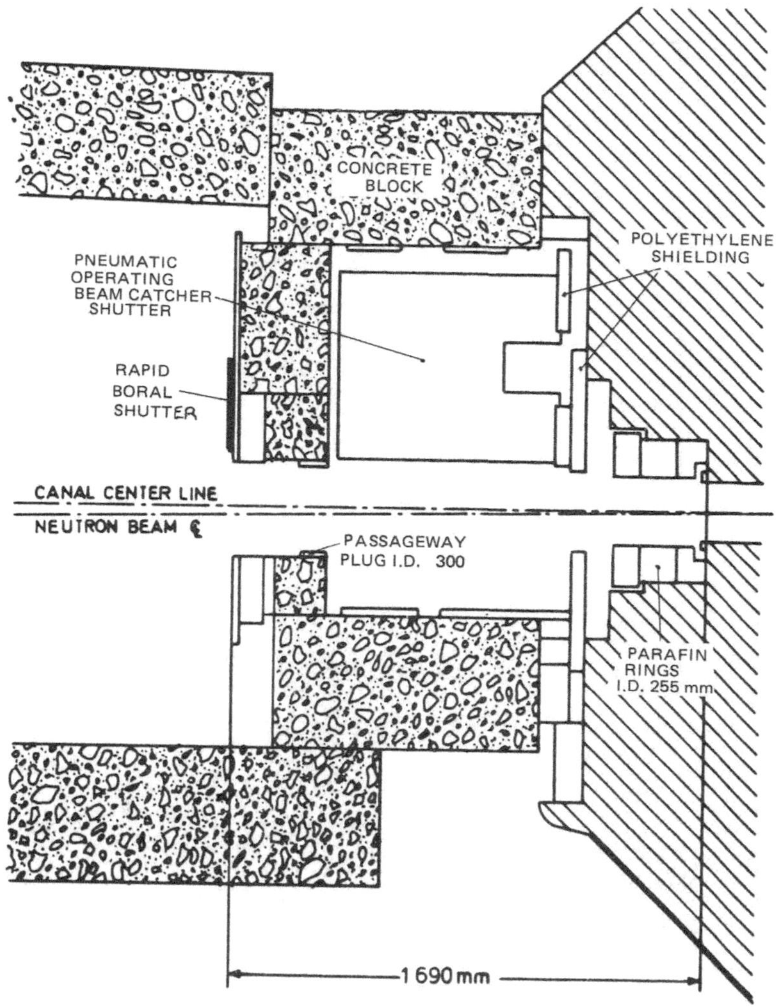

Fig. 5.11 Neutron Radiography Installation at the MELUSINE Reactor, Grenoble Beam Shutter.

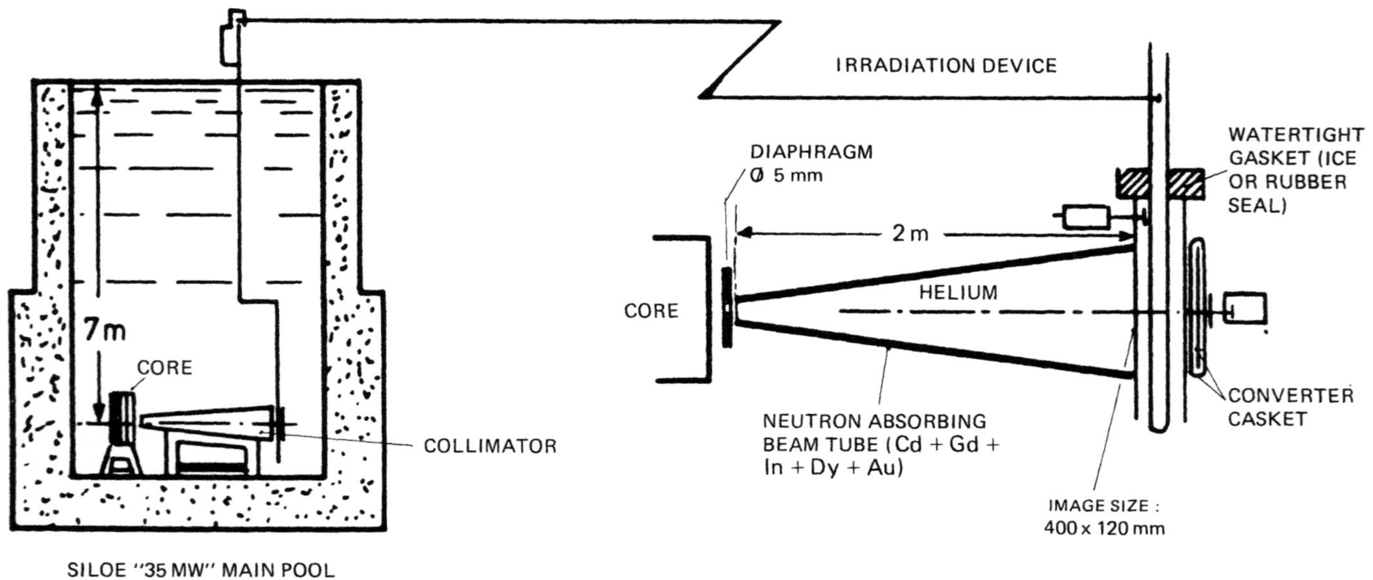

Fig. 5.12 Underwater Neutron Radiography Installation for rigs and loops examination at the SILOE Reactor, Grenoble.

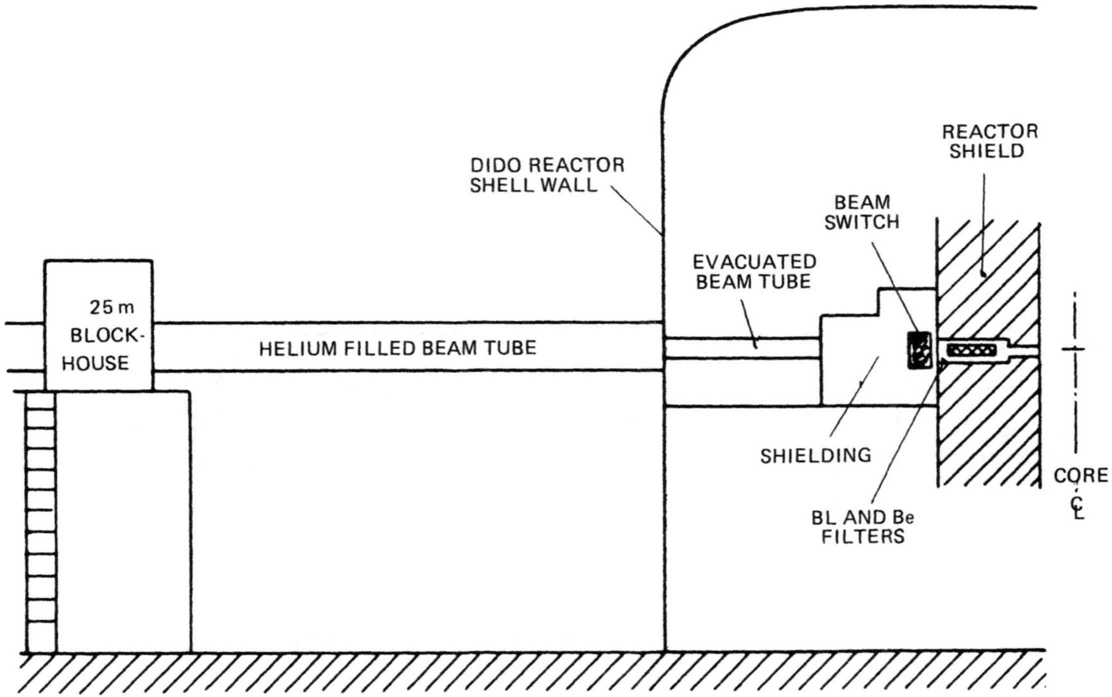

Fig. 5.13 Vertical Section of the 6H Cold Neutron Radiography Apparatus at the DIDO Reactor, AERE Harwell.

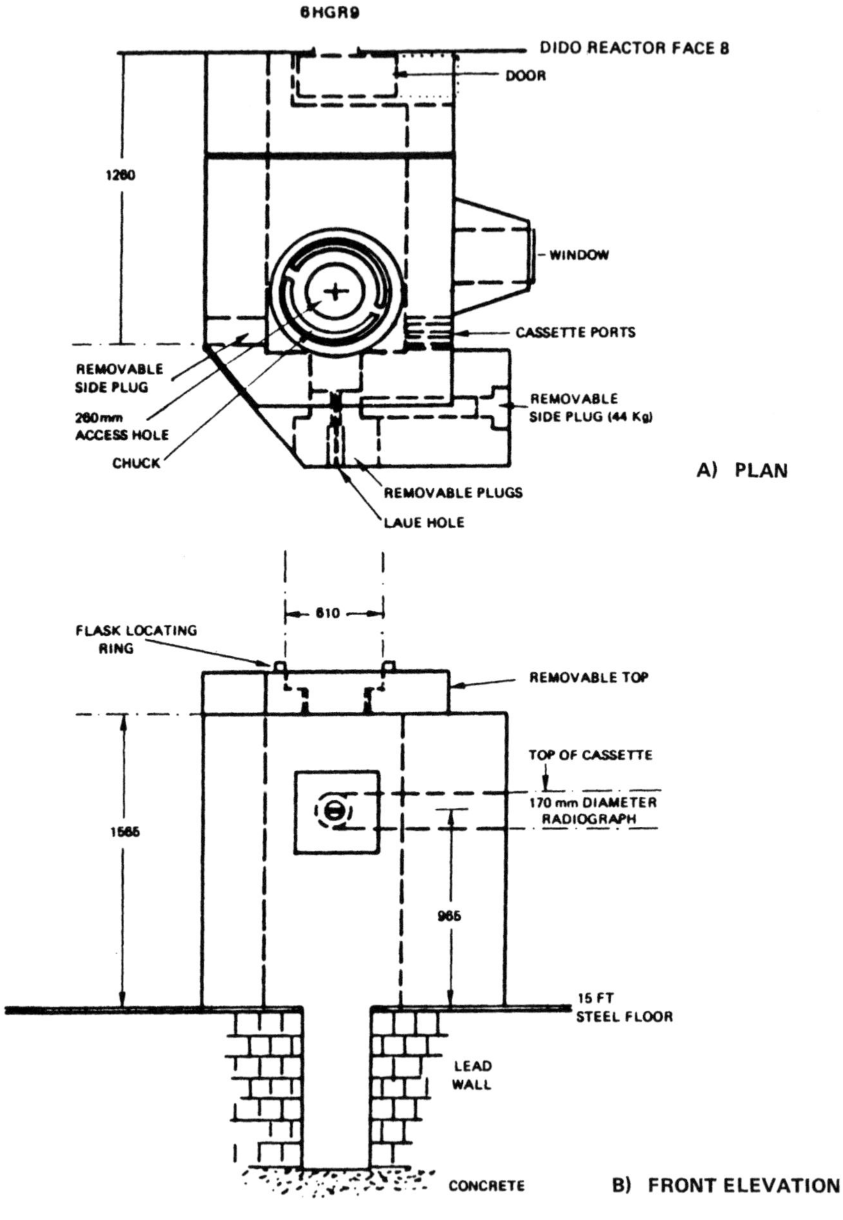

Fig. 5.14 Schematic drawing of the 6HGR9 Neutron Radiography Facility at the DIDO reactor, Harwell.

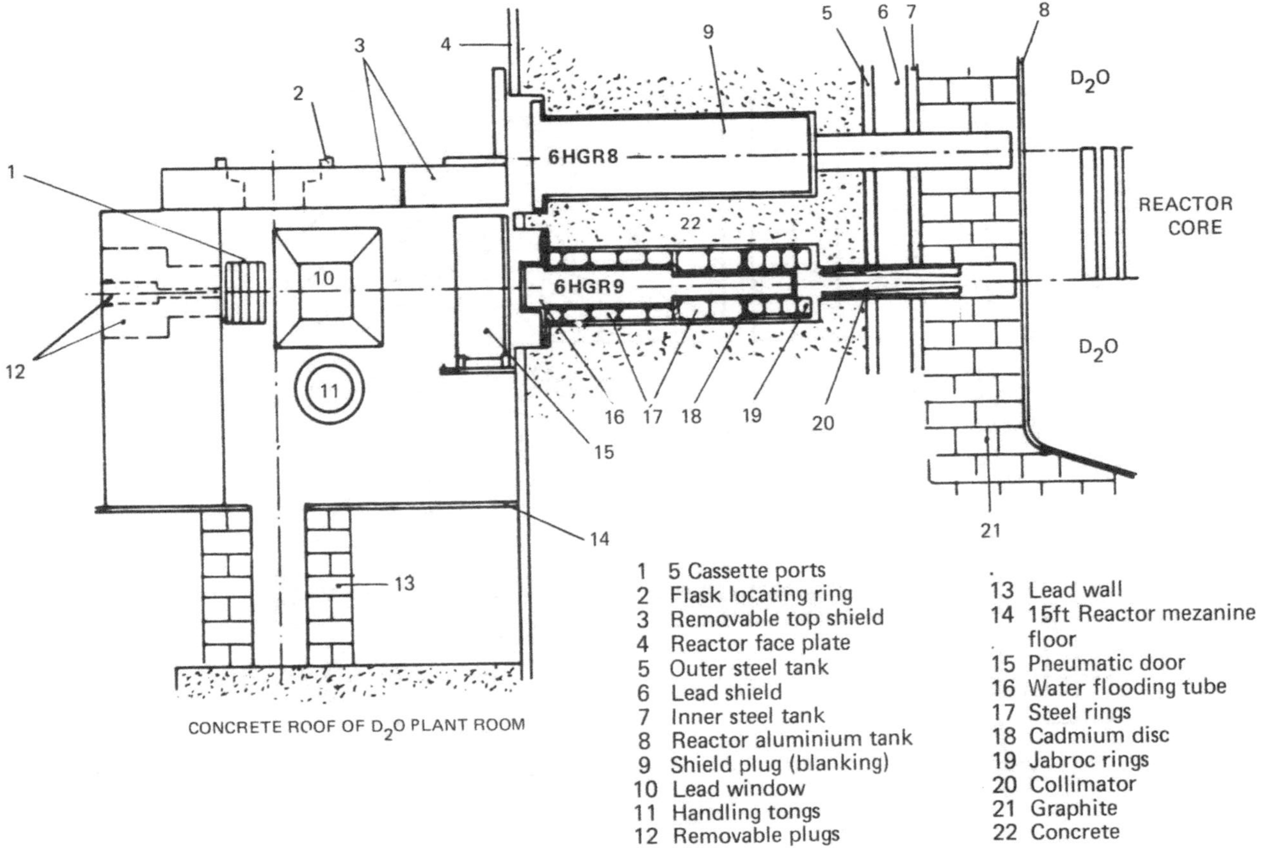

Fig. 5.15 Lay-out of the 6HGR9 Neutron Radiography Installation at the DIDO Reactor, Harwell – Side View.

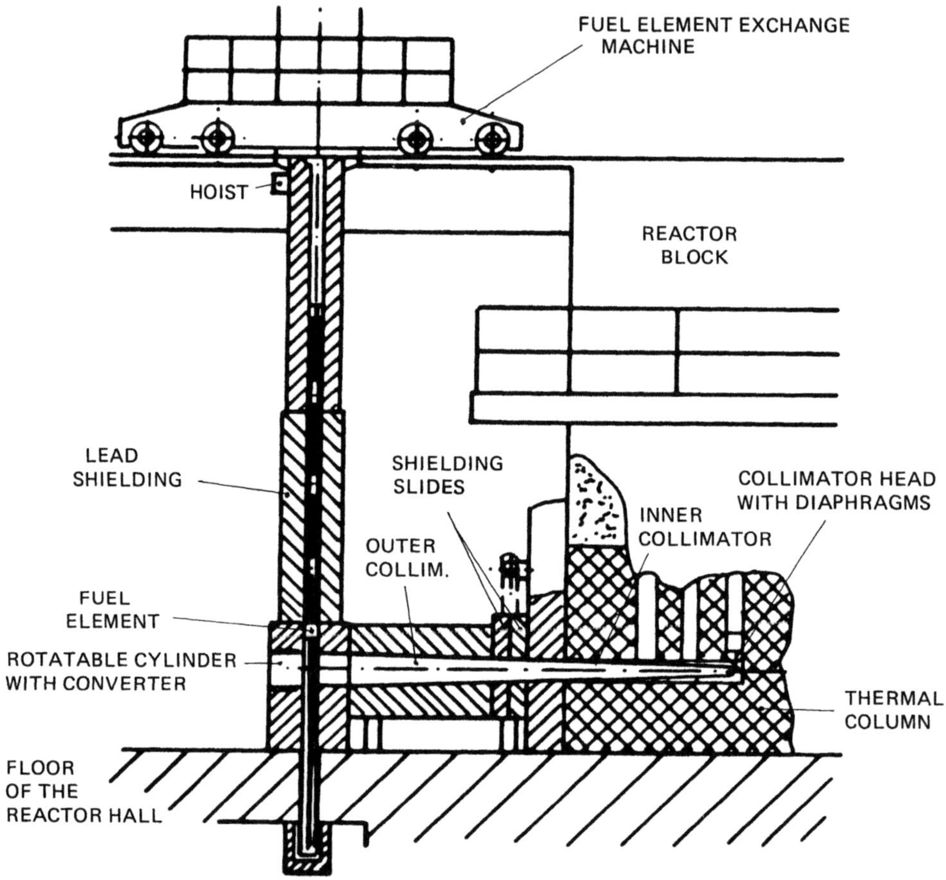

Fig. 5.16 Neutron Radiography Facility for the FR2 at KfK Karlsruhe.

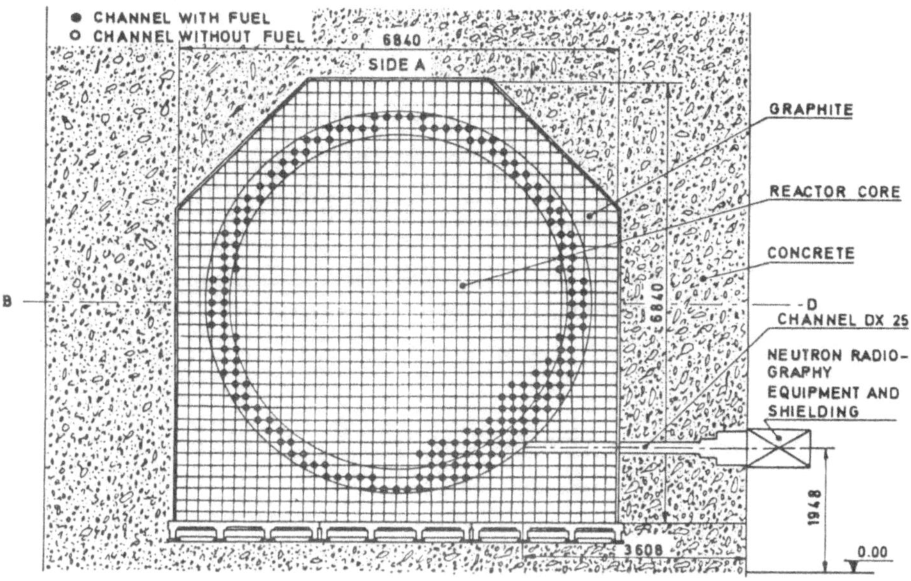

A. GENERAL LAY-OUT

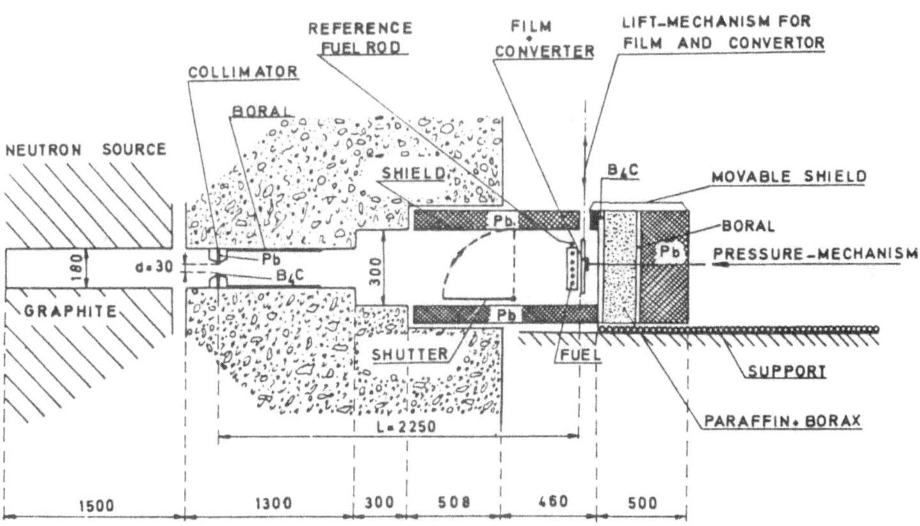

B. DETAILED LAY-OUT

Fig. 5.17 Neutron Radiography Facility at the BR1 Reactor at CEN/SCK, Mol.

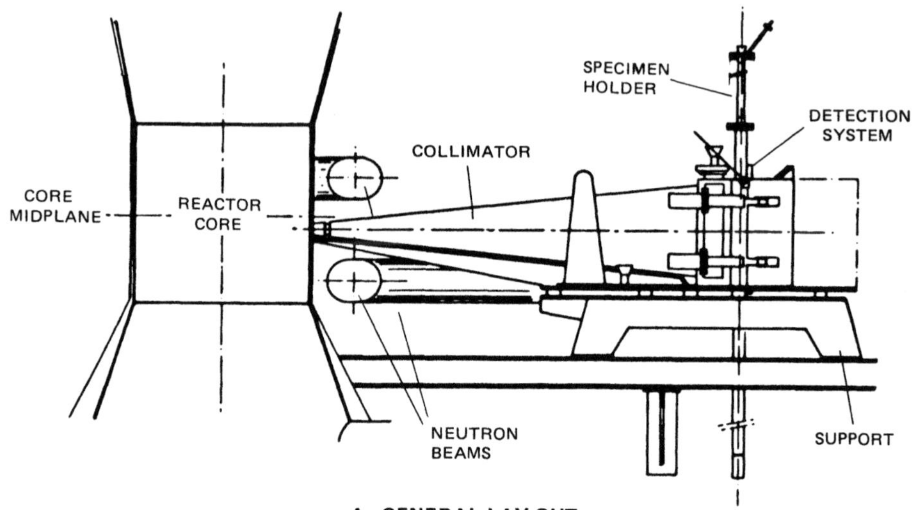

A. GENERAL LAY-OUT

Facility	L_b	L_o
1	2603	28
2	2577,5	53,5
3	2595 – 2655	0 – 60
4	2605	81
5	2615	16

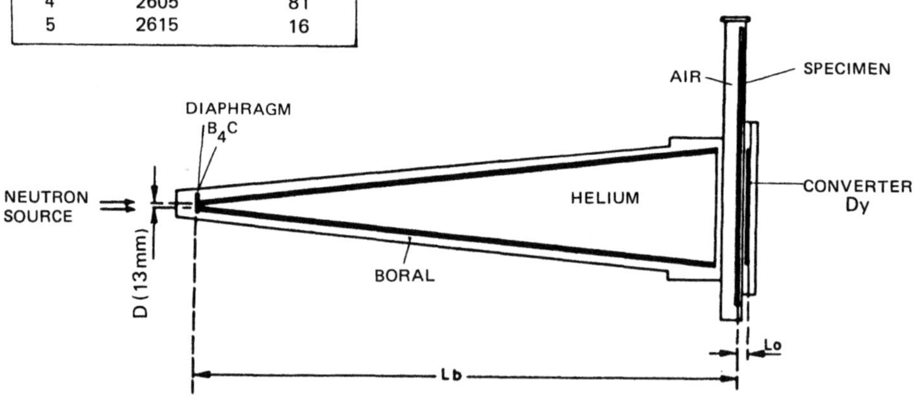

B. DETAILED LAY-OUT

Fig. 5.18 Neutron Radiography Facility at the BR2 Reactor, CEN/SCK, Mol.

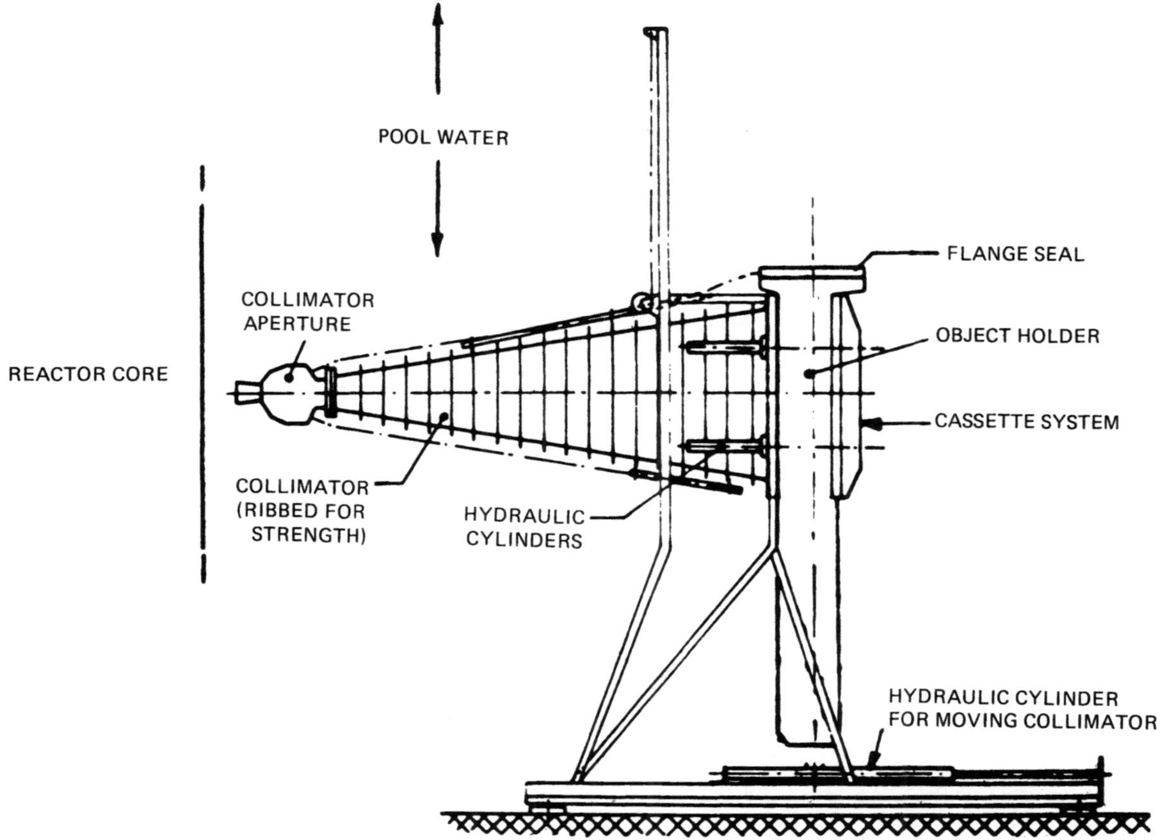

Fig. 5.19 Sketch of the underwater neutron radiography installation in the pool side facility of the High Flux Reactor, HFR, Petten.

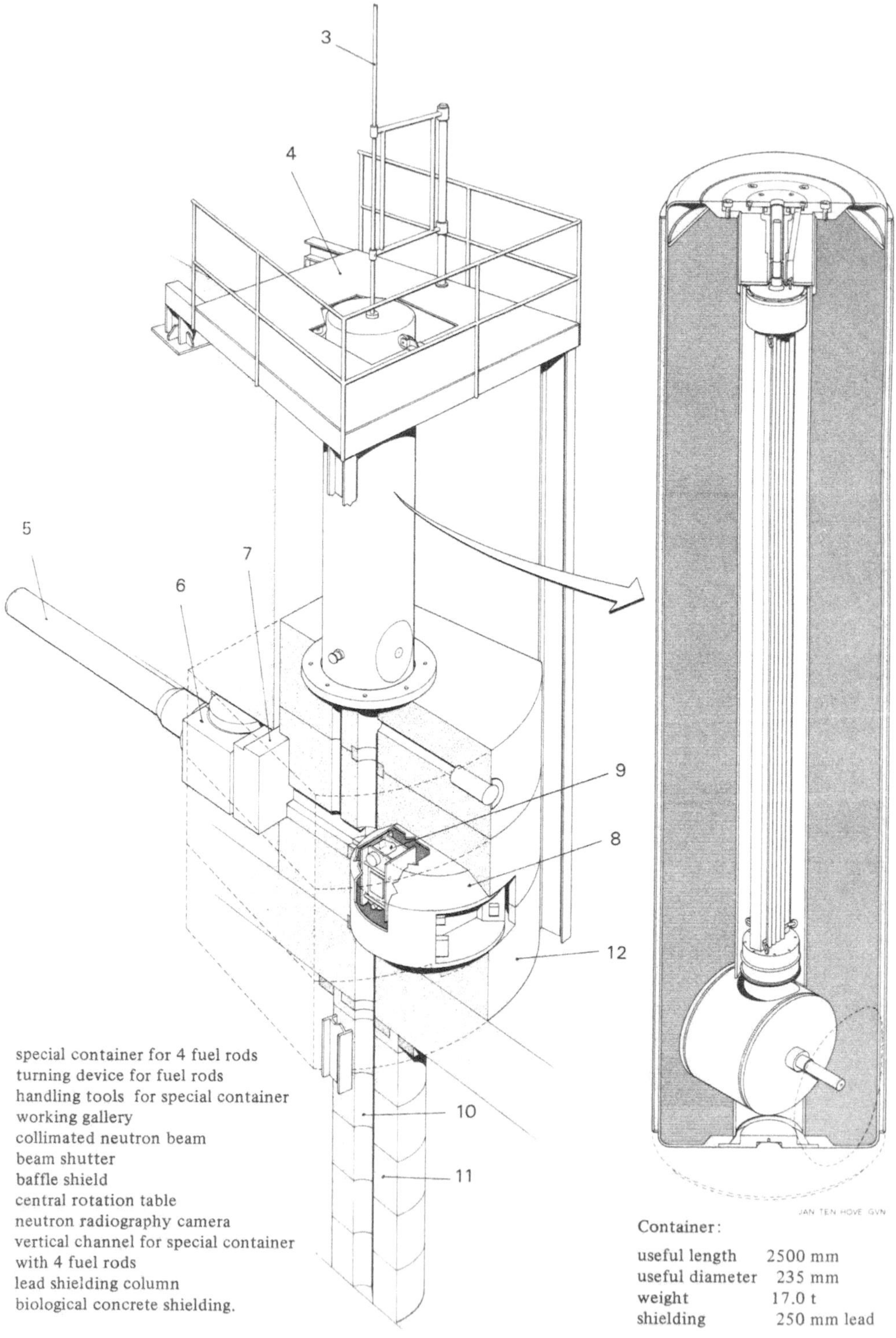

1. special container for 4 fuel rods
2. turning device for fuel rods
3. handling tools for special container
4. working gallery
5. collimated neutron beam
6. beam shutter
7. baffle shield
8. central rotation table
9. neutron radiography camera
10. vertical channel for special container with 4 fuel rods
11. lead shielding column
12. biological concrete shielding.

Container:

useful length	2500 mm
useful diameter	235 mm
weight	17.0 t
shielding	250 mm lead

Fig. 5.20 "Dry" Neutron Radiography Facility HB8 at HFR Petten.

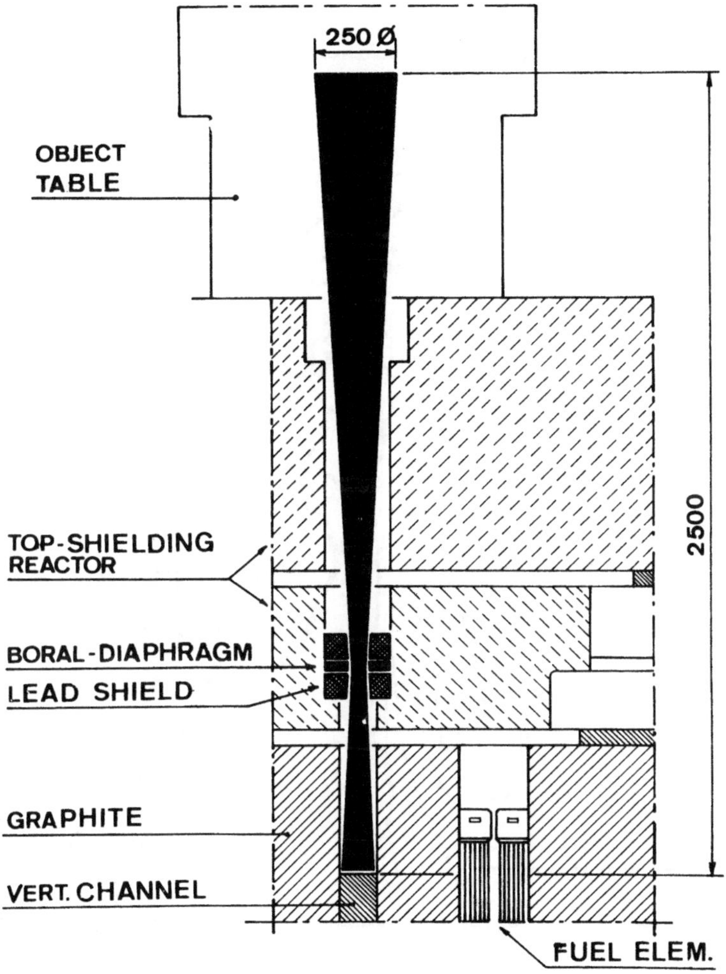

Fig. 5.21 Collimator system of the Neutron Radiography Facility at the Low Flux Reactor (LFR), ECN Petten.

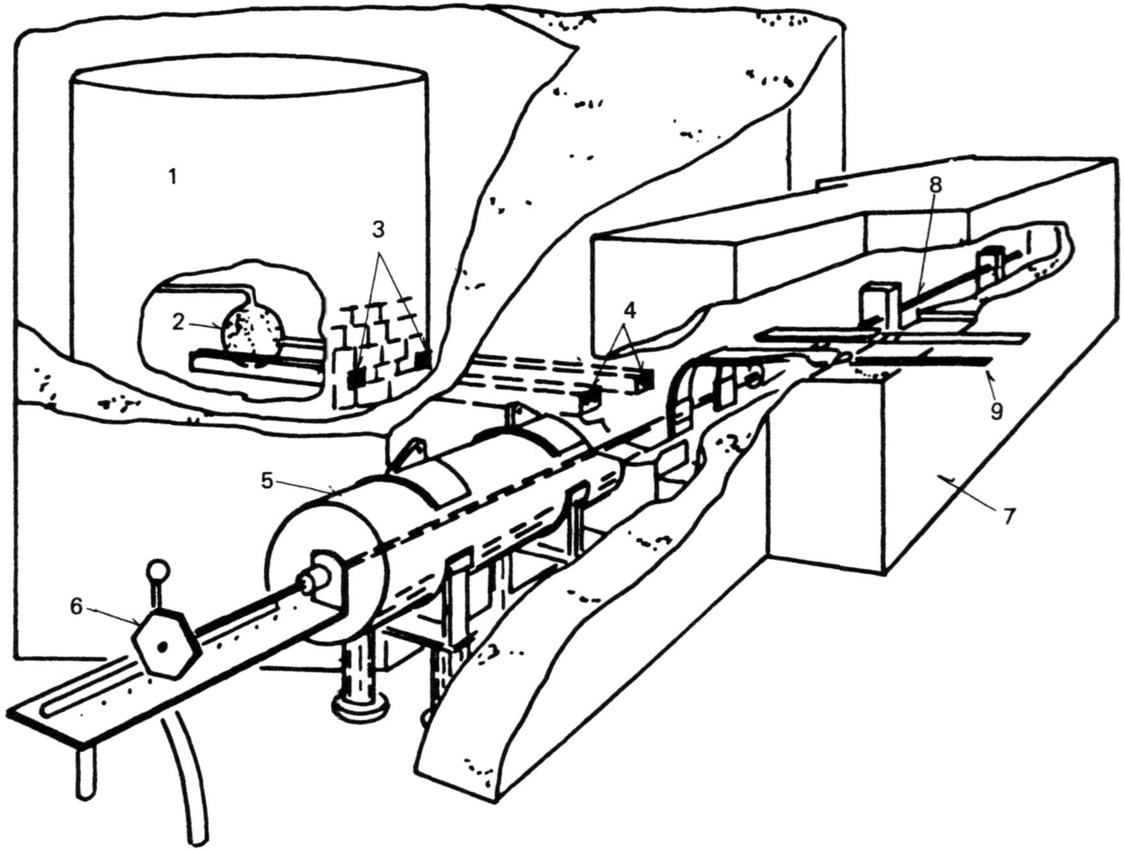

1 Graphite reflector
2 Reactor core
3 Two graphite blocks removed
4 Two neutron beams (10x10 cm)
5 Lead container for transport and handling of irradiated fuel rods
6 Rod for positioning of fuel rod during radiography
7 Concrete blocks for radiation shielding
8 Tube supporting the fuel rod
9 Mechanism for introduction of imaging foils behind the fuel rod to be radiographed

Fig. 5.22 Schematic drawing of the double beam neutron radiography facility at DR1, Risø, Denmark.

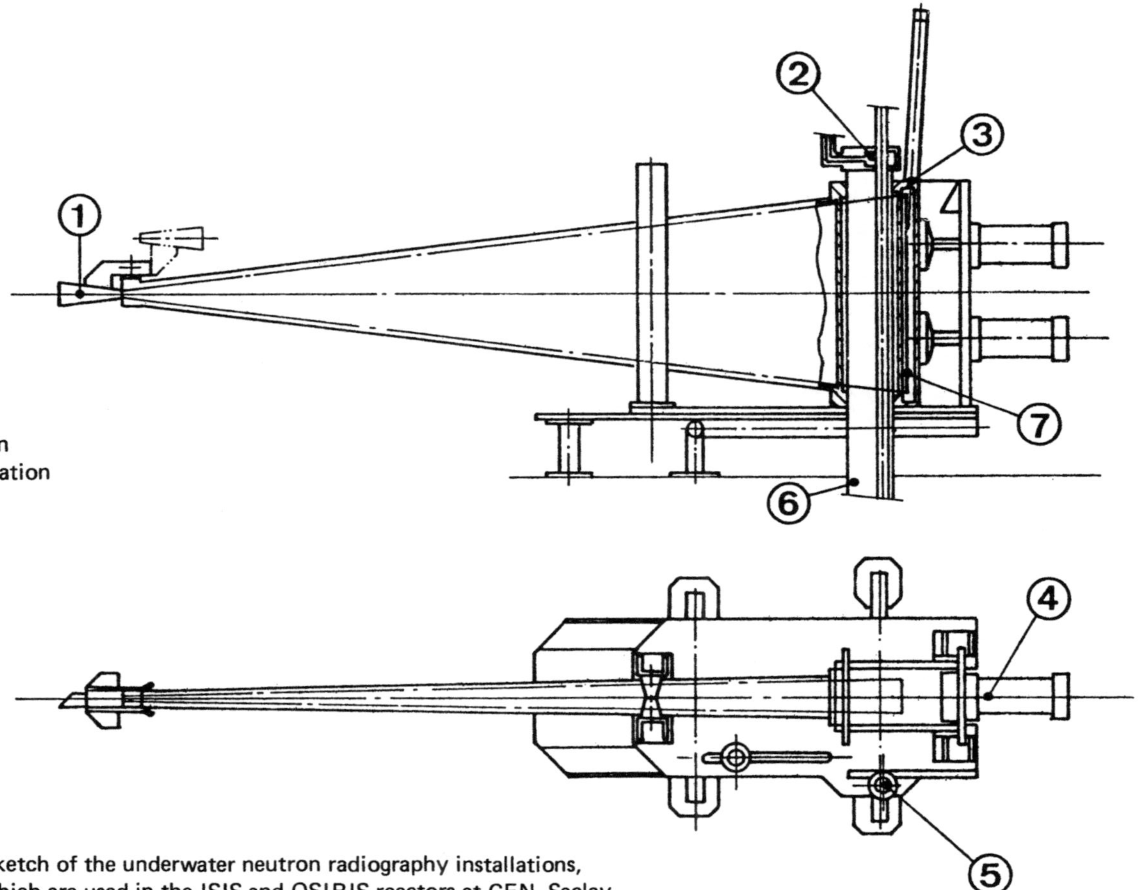

1 Removable precollimation
2 Ice seal from liquid nitrogen
3 Cassette for converter activation and film exposure
4 Removable jacks
5 Displacement control
6 Rig container
7 Converter

Fig. 5.23 Sketch of the underwater neutron radiography installations, which are used in the ISIS and OSIRIS reactors at CEN, Saclay.

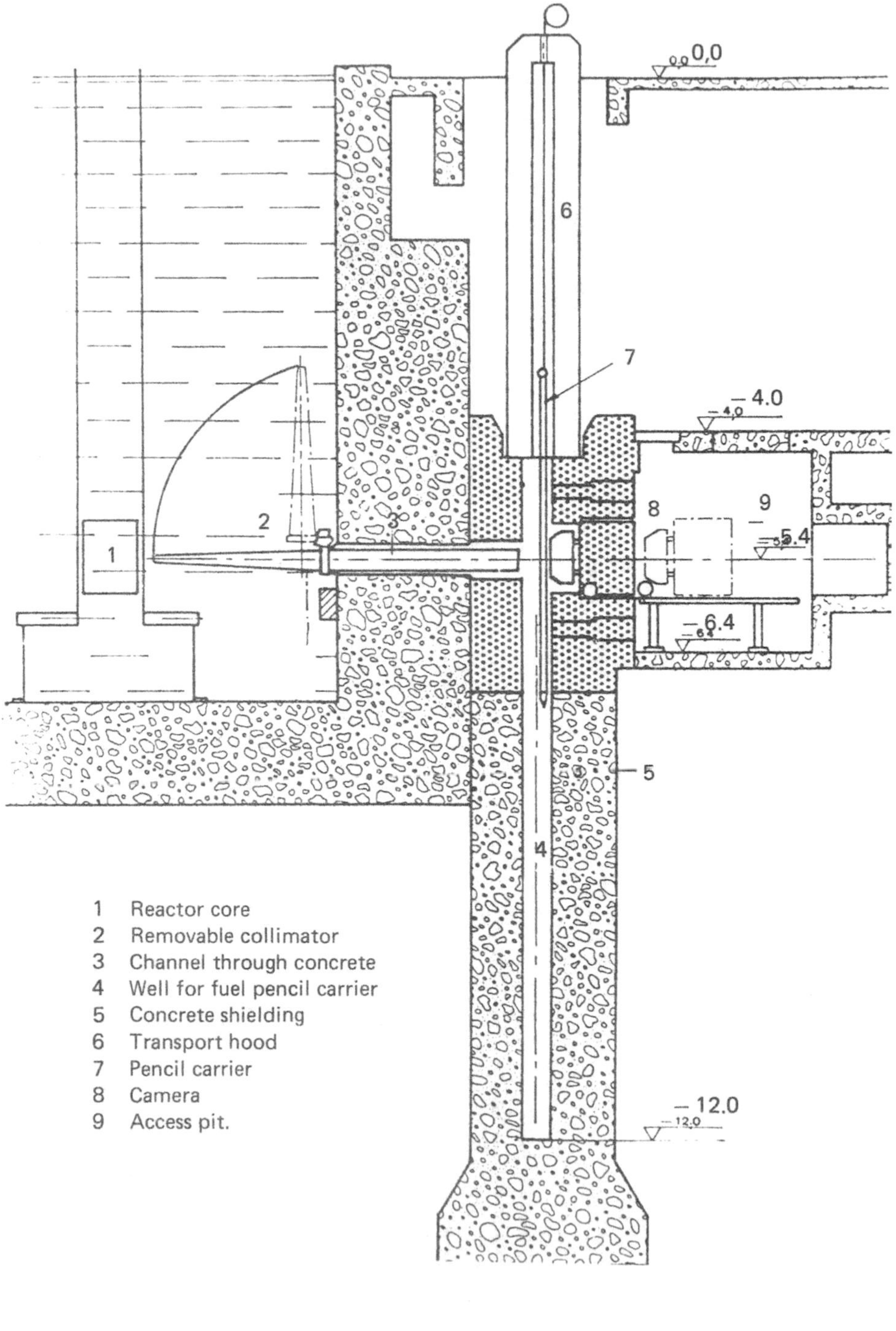

Fig. 5.24 "Dry" neutron radiography facility at ISIS, CEN Saclay.

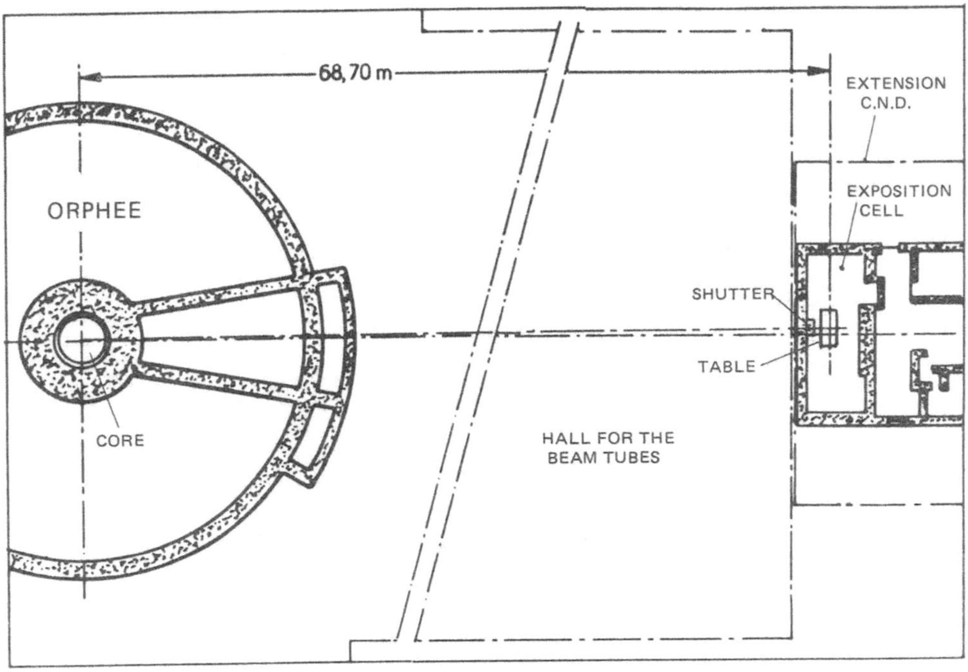

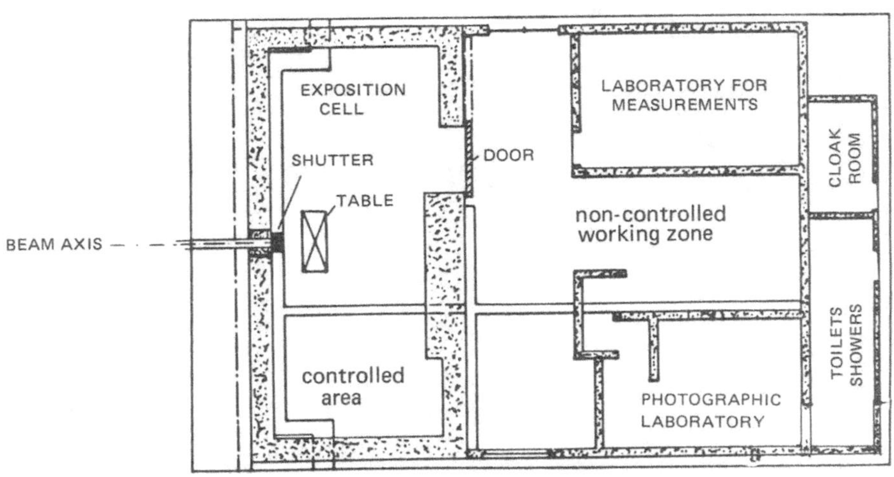

Fig. 5.25 Neutron Radiography Facility at the ORPHEE Reactor, CEN Saclay, General view and Detail of the Exposition Cell and Working Zone.

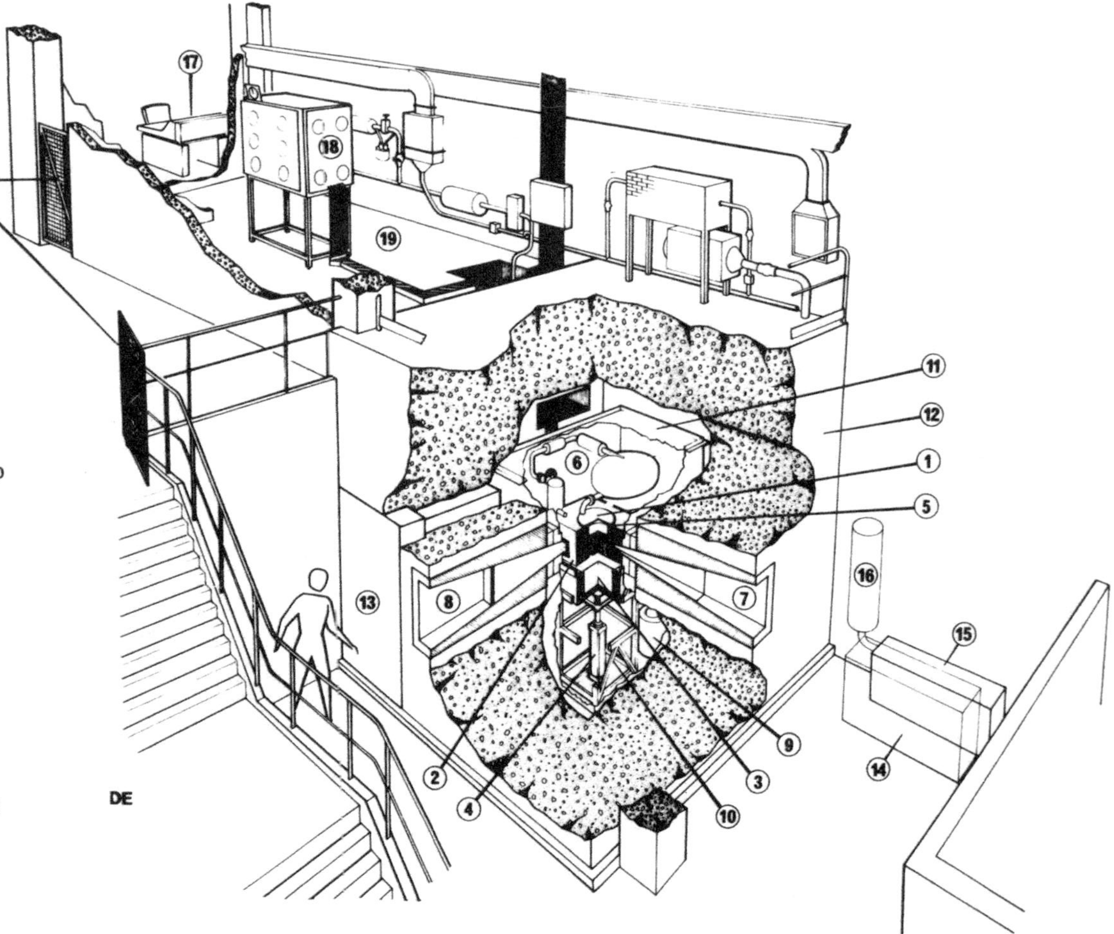

1 Core vessel containing the fissile solution
2 Fixed reflector
3 Mobile reflector
4 Mobile reflector raise - lower cylinder
5 Core cooling or heating loop
6 Recombining loop
7 Axial collimator
8 Tangent collimator
9 Storage tank
10 Frame
11 Caisson
12 Biological shield
13 Inspection door
14 Cold water supply
15 Exchanger
16 Core heating system
17 Control desk
18 Specimen transfer glove box
19 Glove-box compartment
20 Controlled access gates

Fig. 5.26 Schematic drawing of the MIRENE minireactor for Neutron Radiography, Valduc, France.

GPSR Compliance

The European Union's (EU) General Product Safety Regulation (GPSR) is a set of rules that requires consumer products to be safe and our obligations to ensure this.

If you have any concerns about our products, you can contact us on ProductSafety@springernature.com

In case Publisher is established outside the EU, the EU authorized representative is:

Springer Nature Customer Service Center GmbH
Europaplatz 3
69115 Heidelberg, Germany

Batch number: 10607224

Printed by Printforce, the Netherlands